S0-BWX-242

quantitative mass spectrometry in life sciences

quantitative mass spectrometry in life sciences

Proceedings of the First International Symposium held at the State University of Ghent, June 16–18, 1976

editors

a.p. de leenheer

r.r. roncucci

ELSEVIER SCIENTIFIC PUBLISHING COMPANY

Amsterdam – Oxford – New York 1977

ELSEVIER SCIENTIFIC PUBLISHING COMPANY
335 Jan van Galenstraat
P.O. Box 211, Amsterdam, The Netherlands

Distributors for the United States and Canada:

ELSEVIER NORTH-HOLLAND INC.
52, Vanderbilt Avenue
New York, N.Y. 10017

ISBN: 0-444-41557-2

Printed in The Netherlands

CONTENTS

F O R E W O R D

The first International Symposium on Quantitative Mass Spectrometry in Life Sciences was held at the State University of Ghent, Belgium, from June 16th to 18th, 1976. At this meeting 4 plenary lectures and 20 short communications were presented in two days in the Hall of the Akademieraadzaal (Room of the Academic Board) at the State University of Ghent. The 155 registered delegates represented 14 countries (Belgium, Bulgaria, France, Finland, Germany, Italy, Japan, The Netherlands, Rumania, Sweden, Switzerland, Great Britain, U.S.A. and Yugoslavia).

The theme of this First Symposium was devoted to methodological progress and applications of quantitative mass spectrometry. In recent years, the use of quantitative mass spectrometry has revolutionized the measurement of organic compounds, as this technique affords accuracy, specificity and high sensitivity. These are all fundamental to the advancement of the life sciences in general and of bio-analysis in particular. The meeting was organised to provide a forum for the discussion and assessment of the value of recent work in mass fragmentography, and brought together biochemists, clinical chemists, clinicians, organic chemists, pharmacists, pharmacologists, toxicologists, etc., all belonging to disciplines interested in this method.

The plenary lectures presented by E.C. Horning, H. Adlercreutz, C.C. Sweeley and P. Padieu, all leading scientists in this field, are critical surveys of the significance and potential of mass fragmentography in medical, pharmaceutical and biochemical problems. Many of the other contributions also served as excellent examples of progress in quantitative mass spectrometry.

We acknowledge the generous financial support of the Faculty of Pharmaceutical Sciences of the State University of Ghent, the National Foundation for Scientific Research (N.F.W.O.), The Ministry of Public Health and the Ministry of National Education in Belgium. We also are indebted to the following organizations for financial aid: IRE National Institute for Radioelements (Belgium), LKB Instruments N.V. S.A. (Sweden/ Belgium) and Varian-MAT (G.F.R.).

As a result of the succes of the meeting reported in this volume, it is now planned to convene a Second International Symposium on Quantitative

Mass Spectrometry in Life Sciences in September 1978, again in the Ghent area.

We anticipate that one of the merits of this book will be its rapid publication. Usually the proceedings of meetings appear several years after the event. With the cooperation of all of the authors, it has been possible, through the good efforts of Elsevier Scientific Publishing Company, to publish this volume a few months after the meeting.

A.P. DE LEENHEER

R.R. RONCUCCI

QUANTIFICATION OF DRUGS BY MASS SPECTROMETRY

E.C. HORNING, J.-P. THENOT and M.G. HORNING

Institute for Lipid Research, Baylor College of Medicine, Houston, Texas 77030 (U.S.A)

SUMMARY

Current experience indicates that the most generally applicable methods for quantitative analyses of drugs in body fluids are based upon the use of gas chromatograph-mass spectrometer-computer (GC-MS-COM) analytical systems operated in chemical ionization mode with stable-isotope labeled internal standards. Appropriate methods of sample preparation, including derivatization in most instances, should be established for each application.

INTRODUCTION

Mass spectrometers have been used for quantitative analysis for many years, but their present use in biological and medical problems represents a new type of application that requires special study and the development of new technology. The analytical capabilities required today, compared with those of a decade or two ago, are entirely different. Multicomponent analyses of complex mixtures of biologic origin, for compounds present in nanogram or subnanogram amounts, are possible today, but were impossible before the development of current instruments and current techniques. The most useful and most powerful methods now available are based on the use of analytical systems which include three components: a gas chromatograph, a mass spectrometer and a computer. The form of operation is usually in electron impact ionization (EI) mode for identification and structural studies, and in chemical ionization (CI) mode for quantitative analyses. The important events or stages in the development of these systems were the design of "molecule separators" for the concentration of solutes in the gas phase, the use of mass spectrometers as specific ion detectors, the introduction of chemical ionization techniques, and the development of computer-based operation, data acquisition and data analysis capabilities. Along with these mass spectrometric instrumental developments, a parallel set of developments in methods for the derivatization of organic compounds for gas chromatography, and in gas chromatographic columns and instrumentation, have also occurred.

The following sections contain summaries of our current experience with respect to chemical ionization techniques, choice of derivatives, use of stable-isotope labeled internal standards, sample preparation, and precision and accuracy.

RESULTS AND DISCUSSION

Chemical ionization

The basic instrumental technique employed in almost all current quantitative work is that of using a mass spectrometer as a specific ion detector. This technique was described by Holmstedt *et al.*[1] in 1968 in a study of the metabolites of chlorpromazine. The procedure was called "mass fragmentography". The term "selected ion detection" is used more frequently today, but the conceptual basis remains the same. The effluent stream from a gas chromatograph is monitored by a mass spectrometer whose function is essentially that of a detector, and the detection process involves monitoring the appearance of pre-selected ions which are characteristic of the substance(s) under study and the internal reference compounds or internal standards.

The technical problems associated with quantitative studies carried out by this analytical technique involve mass spectrometry (MS) rather than gas chromatography (GC), except for adsorption effects. The GC-MS instrument used by Holmstedt was an LKB 9000, and selective detection was achieved by using rapid cyclic changes in the accelerating voltage, with a fixed magnetic field. This type of operation of a magnetic field GC-MS instrument had been demonstrated earlier by Sweeley, Elliott, Fries and Ryhage[2]. Ion yields were displayed through the usual form of recording on light-sensitive paper. Since these early experiments, numerous changes have been made in the technological aspects. In GC-MS-COM systems of current design, computer-based techniques are used to operate the system and to acquire the necessary data, and data analysis programs are usually also employed. Electrical field instruments are often preferred over magnetic field instruments because of the ease of monitoring ions with relatively large mass differences in the same run, and because of the relative ease of control of operation of electrical field instruments by computers. The major differences in current practices with respect to earlier methods lies in the use of chemical ionization techniques and the use of stable-isotope labeled internal standards.

Chemical ionization methods are currently based on the use of ion sources which are maintained at 0.5-1.0 Torr pressure and in which ionization of sample components occurs by ion molecule reactions. A primary ionization step is carried out by means of electron impact ionization (a heated filament is employed), and a reagent ion is generated from the reagent gas. The most widely used reagent is methane (usually used as the carrier gas as well), but isobutane has also been used, and ammonia and nitric oxide have been used in a number of experimental studies. The reagent ions in these instances are CH_5^+ (methane; some $C_2H_5^+$ and $C_3H_7^+$ are also present); $C_4H_9^+$ (isobutane); NH_4^+ (ammonia; the unreactive ion $NH_3NH_4^+$ is also present); and NO^+ (nitric oxide). The reaction sequence is electrons → reagent ion(s) → sample ions. A separator is not needed; the entire GC effluent stream is directed into the source.

The current period of study of chemical ionization reactions of organic compounds was initiated by the work of Tal'rose *et al.*[3] in 1952, and continued with the work of Field, Franklin and Munson[4] about a decade later. The current "closed" sources came into use in 1969-1971 primarily through the work of Fales, Milne and co-workers[5-7], Munson *et al.*[8] and Biemann *et al.*[9].

A comparison of the EI spectrum of the N,N'-dimethyl derivative of phenobarbital (Fig. 1) with the CI (methane) spectra of the same compound, and the ^{13}C-labeled internal standard (Fig. 2), shows why CI techniques are useful in quantitative work. Only a few product ions are obtained under CI conditions. The chief ionic product corresponds to MH^+; some $(M + 29)^+$ and $(M + 41)^+$ ions are also present, but the cleavage leading to the $(M-28)^+$ ion, which is the major ion formed under EI conditions, is present as a very minor pathway leading to $(MH-28)^+$.

A second advantage of CI techniques in quantitative work is illustrated in Figs. 3 and 4. Fig. 3 shows the EI spectra for the N-methyl derivatives of diphenylhydantoin and a ^{13}C-labeled internal standard. All three isotopic atoms are present in the M^+ ion of the internal standard,

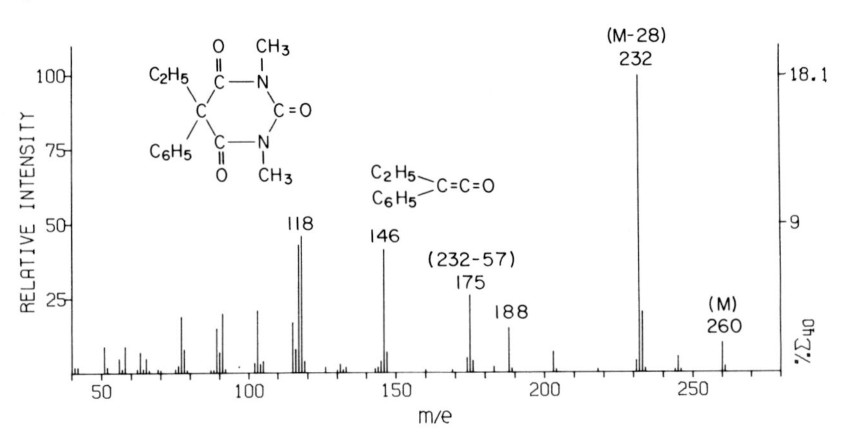

Fig. 1. Electron impact ionization mass spectrum of the N,N'-dimethyl derivative of phenobarbital.

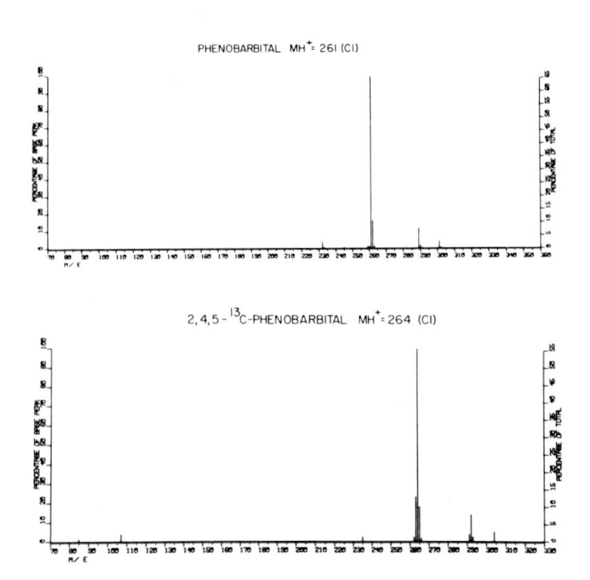

Fig. 2. Chemical ionization (methane) mass spectra of the N,N'-dimethyl derivatives of phenobarbital (upper) and 2,4,5-^{13}C-phenobarbital (lower).

but only one is present in the ion at 181 a.m.u., corresponding to $(M-88)^+$, and this is the major ion in the EI spectrum. The ion at 105 a.m.u. contains only one isotope label; the ions at 239 a.m.u. (M-30) and 211 a.m.u. (M-58) also contain only one isotope label. The CI spectra (Fig. 4), however, show that MH^+ ions are the predominant ions in the spectra, and all isotope labels are retained.

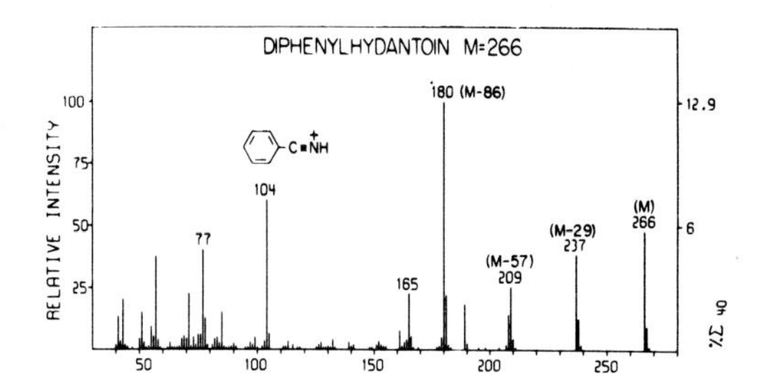

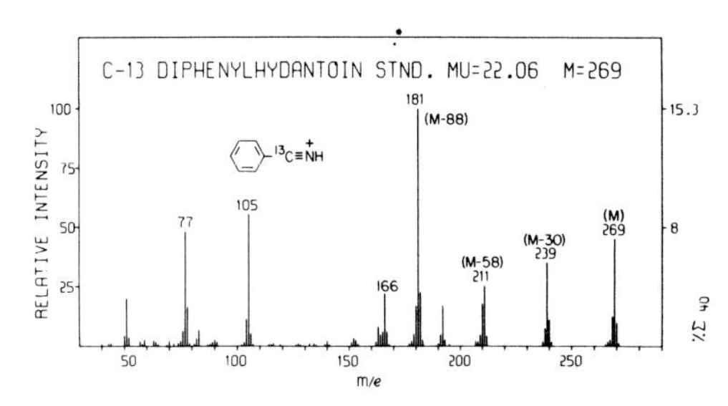

Fig. 3. Electron impact ionization mass spectra of the N-methyl derivatives of diphenylhydantoin (upper) and 2,4,5-^{13}C-diphenylhydantoin (lower). Two of the labeled atoms have been lost in the labeled ion corresponding to the base peak at M-86 (180 a.m.u.) for diphenylhydantoin.

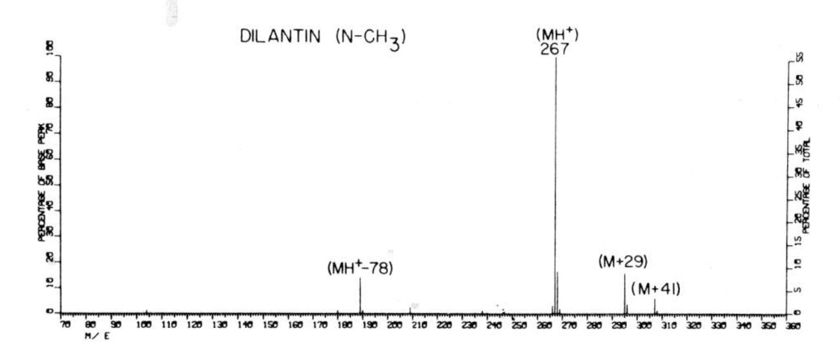

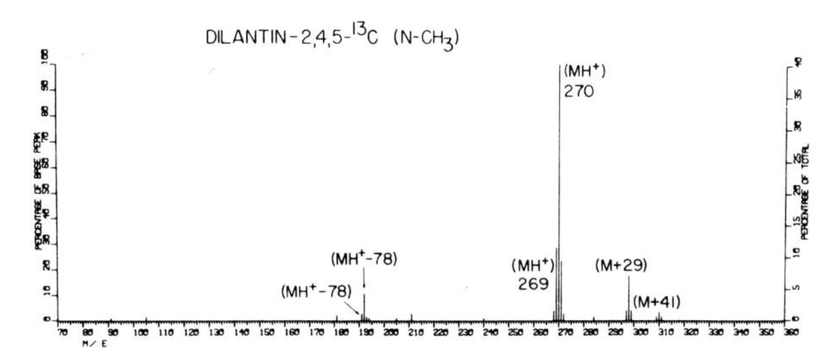

Fig. 4. Chemical ionization (methane) mass spectra of the N-methyl derivatives of diphenylhydantoin and 2,4,5-^{13}C-diphenylhydantoin. The major ions correspond to MH^+.

It is generally believed that methane and isobutane CI spectra always lead to MH^+ ions for compounds containing nitrogen or oxygen, and that the added proton is present on the nitrogen- or oxygen-containing functional group. While this is a useful generalization, no clear evidence has been found for ionic structures containing a protonated trimethylsilyl ether group in an aliphatic or alicyclic structure. The apparent consequence of this protonation reaction is loss of trimethylsilanol, with formation of $(MH-90)^+$ ions. This is not a disadvantage for ^{13}C-labeled internal reference compounds, and usually for ^{2}H-labeled compounds as well.

Continued experience with CI techniques suggests that methane can be used as both carrier and reagent gas in most applications involving packed--column GC separations. Helium is the preferred carrier gas when open tubular capillary columns are used, but nitrogen can also be employed. If two or more ions are present in the CI spectrum, it is often advisable to compare quantitative data based on two ions to make certain that unexpected interference effects are not present.

If CI conditions are used with nitrogen, argon or helium as the carrier gas, the resulting spectra resemble EI spectra taken under 15-20 eV conditions. In nitrogen, for example, at 0.5-1 Torr pressure the principal ions are N^+ and N_2^+, and the energy for ionization by charge transfer is provided by the recombinant energy of these ions. This form of ionization may prove to be useful in some applications.

Choice of derivatives

The chief considerations governing the choice of derivatives for quantitative work are adsorption behavior and stability. In general, the derivatives which have been found most useful are those developed earlier for qualitative studies. Hydroxyl groups are usually derivatized as trimethylsilyl ethers; perfluoroacyl derivatives are used in some applications, and acetyl derivatives have a few specific uses. Acids are usually derivatized as esters (most often as the methyl ester). Ketones are usually converted to methoximes; a few special applications involve other types of ketone reaction products. Primary and secondary amines are usually derivatized as perfluoracyl amides. Tertiary amino groups are usually not derivatized (although they can be converted to substituted carbamates by reaction with chloroformic esters).

Adsorption problems are often encountered. Nicotine, for example, is generally determined without derivative formation, but adsorption effects are severe when very small amounts are involved. In instances of this kind, test procedures must be devised to characterize the behavior of the analytical system near the region of limiting sensitivity of detection for the compounds under study. The use of ^{13}C-isotope labeled compounds is helpful in these instances.

Most derivatives used in gas-phase analytical methods have sufficient stability for quantitative work. Precautions may be needed for relatively labile compounds; for example, N-silyl derivatives are silylating agents, and trimethylsilyl esters of carboxylic acids are easily hydrolyzed. Since the vaporization/sample application zone must be kept in good condition, it is our general practice to use a 1-cm zone of 10% SE-30 column packing at the head of all columns, as recommended by Thenot and Horning[10].

Stable-isotope labeled internal standards

It is now generally recognized that the most satisfactory way of carrying out quantitative analyses for compounds of biologic interest is to employ stable-isotope labeled compounds as internal standards[11]. Compounds containing three or four ^{13}C atoms are preferred. There is no detectable isotope effect in the GC separation process, or in the formation of product ions under CI conditions. Adsorption losses, extraction losses and rates of derivative formation, also show no isotope effect. The introduction of three or four isotopic atoms is desirable in order to avoid overlap with molecular species containing ^{13}C due to the natural occurrence of this isotope. The isotopic composition of internal standards can be inferred from the mass spectrum; for example, Fig. 4 shows the ratio of each type of molecular species present as MH^+ for ordinary diphenylhydantoin (as the N-methyl derivative) and for a ^{13}C-labeled internal reference compound. Confirmatory mass analyses, carried out in the way that response factors are determined, should always be done to ensure that internal reference compounds are chemically pure except for isotope composition variations.

The use of ^{2}H-labeled compounds is frequently necessary because of difficulties in the synthesis of ^{13}C-labeled compounds. For example, it is usually possible to prepare deuterated N-methyl compounds by methylation of a nor-compound with ^{2}H-labeled methylating agents. If deuterium substitution is extensive, alternations will be noted in GC retention behavior; adsorption losses may also be different.

Analogs and homologs are used as internal standards when the appropriate stable-isotope labeled compounds are not available. The least

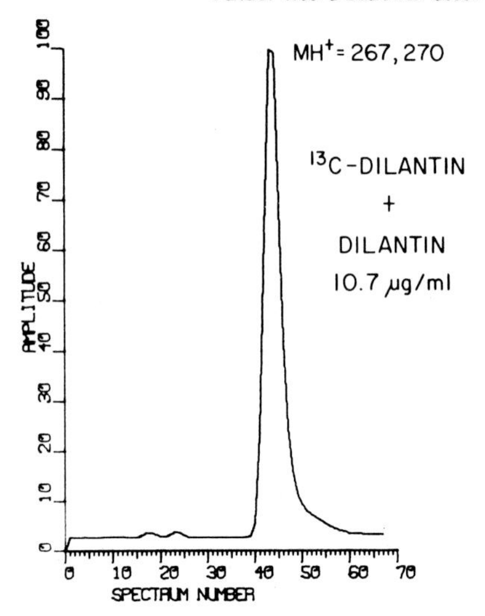

Fig. 5. Selected ion detection (mass fragmentography) chart for the N-methyl derivative of diphenylhydantoin and 2,4,5-^{13}C-diphenylhydantoin. The chromatographic properties are identical. The ions used for detection correspond to MH^+.

satisfactory circumstance is to use an unrelated compound as an internal standard under EI conditions. In theory, this would be possible if measured ion ratios were always constant for different compounds, with present methods of control of ion source conditions. In practice, measured ion ratios for unrelated types of compounds may not be sufficiently reproducible.

Fig. 5 shows a computer-plotted chart of the ion current observed during an analysis of a biologic sample (derived from human plasma) containing an added ^{13}C-labeled internal standard. Ions corresponding to MH^+ for the N-methyl derivatives were monitored at 267 and 270 a.m.u. The retention behavior of the labeled and unlabeled compounds was identical, as expected. Programmed analysis of the data indicated that the original concentration of diphenylhydantoin (phenytoin) in the plasma sample was 10.7 μg/ml.

Measured ion ratios may be computed on the basis of either peak height or peak area comparisons. When ^{13}C-labeled compounds are used as internal standards, peak height comparisons are usually satisfactory. In other instances, peak area comparisons are usually employed. Fig. 6 shows

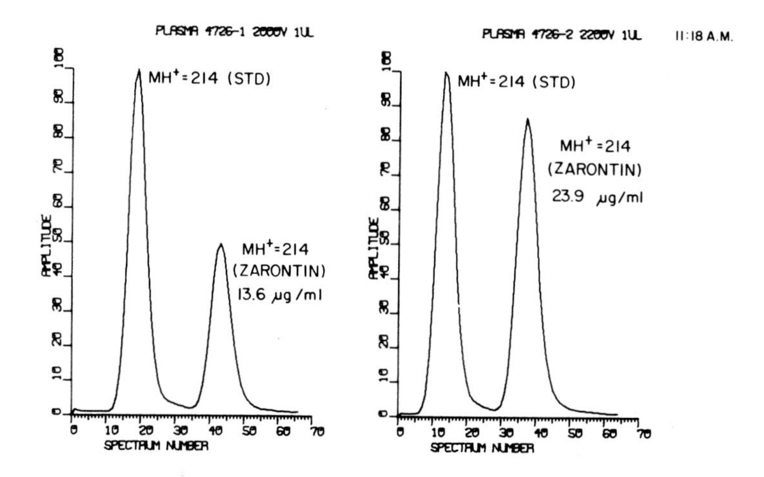

Fig. 6. Selected ion detection (mass fragmentography) chart for the N-trimethylsilyl derivatives of 1-methyl-1-ethylsuccinimide (Zarontin) and 1,1,2-trimethylsuccinimide (internal standard). The ions used for detection correspond to MH^+.

two examples of human plasma analyses for 1-methyl-1-ethylsuccinimide as the N-trimethylsilyl derivative, with 1,1,2-trimethylsuccinimide as the internal standard. In this instance, the same a.m.u. value is monitored, corresponding to MH^+ for both compounds, but the retention times of the isomers are different. Peak area comparisons were made by programmed analysis of the data.

Sample preparation

The basic steps in sample preparation procedures are usually the same as those used in identification or detection studies, with modifications as required for each problem. In some instances, procedures can be simplified. For example, when ^{13}C-labeled internal standards are used, it is usually adequate to carry out only one organic solvent extraction of a biologic fluid for a drug or a drug metabolite. Several extractions will increase the amount of sample, but the ratio of compound to internal standard will remain the same whether the extraction process involves one, two or three extractions. Purification of the sample is frequently needed, however, and this is critical when varying degrees of interference are encountered in biologic studies. For example, the analysis of total morphine in human urine is carried out after an enzymic hydrolysis step which results in the liberation of urinary steroids. The morphine/steroid mixture obtained

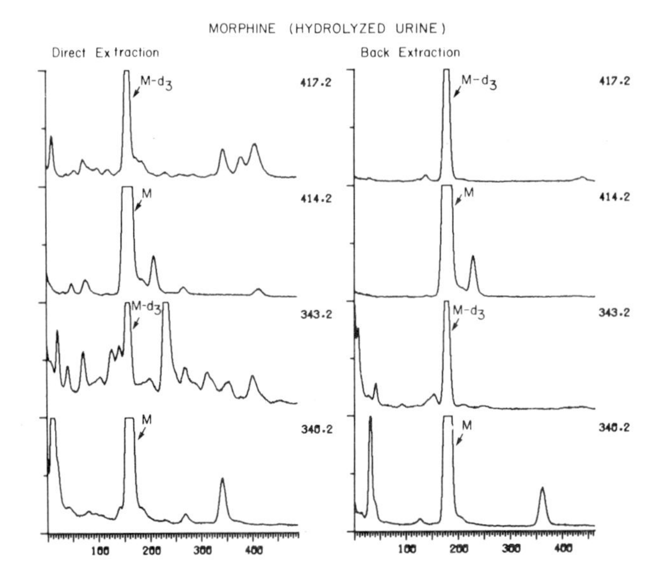

Fig. 7. Selected ion detection (mass fragmentography) charts for the estimation of total morphine in urine. The derivatives are the O^3,O^6-trimethylsilyl ethers of morphine and of morphine-d_3. The ions correspond to $(MH-16)^+$ and $(MH-90)^+$ under chemical ionization (methane) conditions. The sample as left was obtained by direct extraction after enzymic hydrolysis (Glusulase); the sample at right was purified by back extraction. The interfering substances are urinary steroids.

after organic solvent extraction is not suitable for direct quantitative analysis because of interference with the internal standard (morphine-d_3) measurements. A back extraction procedure (or an ion exchange purification) which removes the neutral steroids results in a satisfactory sample. This effect is illustrated in Fig. 7.

Precision and accuracy

The precision of quantitative analyses of biologic samples carried out with a GC-MS-COM analytical system, with stable-isotope labeled internal standards, should be about 2-4% for the instrumental analysis step. This is illustrated in Table I for analyses for phenobarbital and diphenylhydantoin, carried out with a Finnigan GC-MS-COM analytical system operated in CI mode with methane as the carrier and reagent gas, and with monitoring for MH^+ ions (N-methyl derivatives). When two ions resulting from cleavage reactions are monitored, it is possible to confirm compound ratios and to estimate the precision for each measurement. Table II shows

Table I

Multiple quantitative analysis of a plasma sample containing phenobarbital (PB) and diphenylhydantoin (DPH)

Run		PB(μg/ml)	Δ		DPH(μg/ml)	Δ
1		38.6	0.5		17.6	0.1
2		37.9	0.2		17.4	0.1
3		38.5	0.4		17.8	0.3
4		37.5	0.6		17.1	0.4
5		37.0	1.1		17.5	-
6		40.4	2.3		18.8	1.3
7		40.4	2.3		17.5	-
8		39.1	1.0		17.4	0.1
9		35.1	3.0		18.0	0.5
10		36.8	1.3		16.1	1.4
	Ave.	38.1	1.3	Ave.	17.5	0.4
	Std.dev.		3.3%			2.4%

data for successive analyses of a urinary morphine sample with a Finnigan system, using methane as the carrier and reagent gas, and morphine-d_3 as an internal standard, with monitoring of ions corresponding to $(MH-16)^+$ and $(MH-90)^+$ (O^3,O^6-trimethylsilyl derivatives). The relative standard deviation of ion count ratios for these two ions was less than 2%.

Table II

Analyses of a urinary morphine sample as its trimethylsilyl derivative

		$\frac{414}{417}$	$\frac{340}{343}$
A		1.313	1.268
B		1.339	1.292
C		1.309	1.257
D		1.310	1.256
E		1.286	1.225
	Mean	1.311	1.260
	RSD	1.4%	1.9%

When difficulties are encountered in establishing satisfactory precision in the instrumental analysis step, it is usually necessary to examine a number of possible sources of error. In instruments of older design, drift is a frequent source of difficulty. Ordinary computer--based operation procedures may be unsatisfactory when mass defects are moderately large and calibrations are not sufficiently good; for the work in Table II, the mass spectrometer was calibrated with the compound under study. Varying column losses may lead to poor precision when suitable stable isotope labeled compounds are not available, and when very small samples are employed.

Repetitive analyses of the same biologic sample generally show slightly larger standard deviations. It is desirable to compare peak height and peak area measurements to determine if unexpected interference is present; when ^{13}C-labeled compounds are used as internal standards, the values should be the same, and the precision should be the same. This is illustrated in Table III, for plasma analyses for phenobarbital and diphenylhydantoin.

Table III

Analysis of quadruplicate extracts of plasma by SID

Sample		Phenobarbital		Dilantin	
		Peak height (μg/ml)	Peak area (μg/ml)	Peak height (μg/ml)	Peak area (μg/ml)
3A		39.4	40.0	9.0	9.3
3B		38.3	39.3	8.2	8.3
3C		38.6	37.0	8.7	7.9
3D		39.8	39.9	8.1	8.1
	Average	39.0 ± 0.4	39.0 ± 0.75	8.5 ± 0.2	8.4 ± 0.35

The accuracy of an analytical method for a biologic sample is generally defined by the use of spiked samples, and quality control is maintained by interposing reference samples in each series of analyses.

It is not possible to define the accuracy to be expected from GC-MS--COM methods in all applications, but in most work the accuracy should approach the precision. This assumes that appropriate precautions are taken in sample preparation steps, and that sources of varying degrees of interference or loss have been identified and eliminated. (This is usually possible at an early stage in methods development.)

ACKNOWLEDGEMENT

The work described here was supported in part by Grants GM-13901 and GM-16216 of the National Institute of General Medical Sciences and Contract DAMD-17-74-C-4052.

REFERENCES

1 C.-G. Hammar, B. Holmstedt and R. Ryhage, *Anal. Biochem.*, 25 (1968) 532.
2 C.C. Sweeley, W.H. Elliott, I. Fries and R. Ryhage, *Anal. Chem.*, 38 (1966) 1549.
3 V.L. Tal'rose and A.K. Lyubimova, *Dokl. Akad. Nauk S.S.S.R.*, 86 (1952) 909.
4 F.H. Field, J.L. Franklin and M.S.B. Munson, *J. Amer. Chem. Soc.*, 85 (1963) 3575.
5 H.M. Fales and G.W.A. Milne, *J. Amer. Chem. Soc.*, 91 (1969) 3682.
6 H.M. Fales, H.A. Lloyd and G.W.A. Milne, *J. Amer. Chem. Soc.*, 92 (1970) 1590.
7 D. Beggs, M.L. Vestal, H.M. Fales and G.W.A. Milne, *Rev. Sci. Instr.*, 42 (1971) 1578.
8 D.M. Schoengold and B. Munson, *Anal. Chem.*, 42 (1970) 1811.
9 G.P. Arsenault, J.J. Dolhun and K. Biemann, *Chem. Commun.*, (1970) 1542.
10 J.-P. Thenot and E.C. Horning, *Anal. Lett.*, 5 (1972) 801.
11 M.G. Horning, J. Nowlin, K. Lertratanangkoon, R.N. Stillwell, W.G. Stillwell and R.M. Hill, *Clin. Chem.*, 19 (1973) 845.

QUANTITATIVE MASS SPECTROMETRY OF ENDOGENOUS AND EXOGENOUS STEROIDS IN METABOLIC STUDIES IN MAN

H. ADLERCREUTZ

Department of Clinical Chemistry, University of Helsinki, Meilahti Hospital, Helsinki (Finland)

SUMMARY

A review on our experiences with quantitative mass spectrometry of endogenous and exogenous steroids in metabolic studies in man is presented. Endogenous estrogens have been studied in urine and bile of pregnant and non-pregnant subjects and in plasma and feces from pregnant subjects. Metabolic studies were also carried out with megestrol acetate, estriol and 16α-hydroxyestrone after oral administration.

INTRODUCTION

Mass fragmentography or selected-ion monitoring of steroids was described before 1970[1], but only few publications containing information on this subject had appeared by 1971[2-7]. The development of this field[8,9] has been partly overshadowed by the extremely rapid progress made with radioimmunological and related techniques, which are simpler and permit the analysis of large numbers of samples. Quantitative mass spectrometry is a technique that still requires special skill and experience; therefore metabolic studies of steroid hormones using this technique are carried out only in a limited number of laboratories. There is no doubt but that both radioimmunological and mass fragmentographic methods have contributed significantly to our understanding of steroid hormone metabolism, and these techniques must be regarded as complementary.

GENERAL CONSIDERATIONS

Table I outlines the main reasons for using mass fragmentography in the studies that will be described below. Table II displays the main reasons we have had for choosing not to use selected-ion monitoring in several other studies. It is a great advantage to have a choice of several techniques, the main problem being to keep them all going.

Table I

Main reasons for using mass fragmentography instead of other techniques in metabolic studies

1. Possibility of simultaneously measuring structurally related steroids (*e.g.* epimeric steroids)
2. Novel biological materials of unknown composition (*e.g.* portal blood) may be assayed even if only small samples are available
3. Superior specificity as compared with GC and RIA
4. Determination is possible of small amounts of estrogens in samples containing high concentrations of neutral steroids (*e.g.* feces)
5. Further specificity studies possible with the rest of the sample

Table II

Main reasons for not using mass fragmentography in metabolic studies

1. Large numbers of samples
2. Only 1-2 steroids to be assayed and specific antibody available
3. GC preferred if steroid concentration is high and many steroids must be assayed
4. Better precision of RIA and GC assays
5. If sensitivity of the mass fragmentographic method is not sufficient (RIA or enzymic techniques better)

In most publications on the quantitative mass spectrometry of steroids the reliability of the methods receives only little discussion, although in practice one encounters many problems. It seems clear that much more attention should be paid to the reliability of these methods. This problem should also be very carefully considered by the manufacturers of mass spectrometers, as some of the variation in results is due to instrumental defects. Our experiences in this field are based on the utilization of four different instruments. 1. An LKB 9000 with an accelelerating voltage alternator (AVA). As this is our own instrument we have had most experience with it. It belongs to the first series and is more than 10 years old. 2. A Varian MAT CH7 with a peak matching system and equipment for recording two ions by using an accelerating voltage alternator. 3. An LKB 2091 instrument with the new MID system. 4. A Varian MAT 112. In the sensitivity experiments single-ion monitoring was used, and in the experiments for evaluating precision we used two ions, one for the estrogen under investigation and one for the internal standard, which was always prepared with deuterated silylating reagents and resulted in the internal standard having a 9, 18 or 27 a.m.u. higher mass than the estrogens investigated, depending on its hydroxyl group content. All precision experiments were carried out with 100-200 pg of steroid.

When testing the *sensitivity* of the instruments we used packed columns, the test substances being methylated and silylated estradiol and silylated estriol. We found the 3-methoxyestradiol-17-trimethylsilyl ether the most suitable compound for such tests as the molecular ion at *m/e* 358 gave a very sensitive measurement of this steroid. For all the instruments mentioned it was possible to work with steroid amounts below 10 pg. The smallest amount giving a peak-to-noise ratio of about 3-5 was 2.5-3 pg. This level of sensitivity was obtained both with the LKB 2091 and the Varian CH7. With the other two instruments similar results were obtained with about 5-8 pg; so, in practice, there was no significant difference between the instruments as the conditions of operation may have varied to some extent despite careful control. Differences may have arisen owing to variation in the quality of the gas chromatographic columns. A picture of the recording of a full-scale response with 50 pg estriol-TMS from pregnancy urine has previously been published[10].

When measurements are carried out at the nanogram level, reasonable assay *specificity* and *accuracy* can usually be achieved without difficulty. No other technique can give comparable specificity at that level. Below the nanogram level other compounds in the biological extracts may interfere;

therefore measurements of several characteristic fragment ions are recommended. With the old AVA system serious problems occurred, when internal standards with the same or nearly the same retention time as that of the analytical material were used. Under such conditions a large amount of substance being measured with one channel will increase the response on the other channel in an unpredictable way. If a dynamic study is being undertaken and the level of endogenous steroid varies greatly the measurements can, unfortunately, be completely unreliable, as the recording of a high peak in one channel will prevent the recorder from reaching the baseline and thereby increase the measured amount on the second channel. Thus, if the old AVA system is used, the internal standard must have a retention time clearly different from that of the analytical material or else the levels of both must be about the same. Such problems do not occur with the new MID systems.

In the presentation of mass fragmentographic methods it is essential to give details of their *precision*, as this very much depends on the amount injected, on the stability of the instrument, on the gas chromatographic column and on some other factors. In 1975 Lee and Millard[11] reported that satisfactory precision at the 100-pg level can only be achieved after extensive overhaul of the instrument. They showed that multiple-ion monitoring is less precise than single-ion monitoring and that coefficients of variation of about 15% are common even at levels of 1-10 ng. Some of our results are shown in Table III.

Table III

Results of precision studies with reference estrogen standards (100-200 pg) using various instruments

Instrument	Mean coefficient of variation (%)	Number of experiments
LKB 9000 + AVA	8.9	5
LKB 2091 + MID	8.0[+]	6
Varian MAT CH7 with peak matching system	1.2[++]	1
Varian MAT 112 + MID	6.0	12

[+] One very high value excluded.
[++] Duration of experiment 36 h.

The results presented in Table III were obtained with reference standards, which, in each experiment, were injected at least 5 times. It should be emphasized that we varied the conditions during the experiments, changed the amount of internal standard and analytical material in the same proportion, and altered the amplification of the signal to a different extent for each channel. Experiments were carried out over a whole day and were interspersed between analyses of bile samples to simulate typical working conditions during the investigations. The instruments were not recalibrated during the working period.

We found two factors in particular that caused a decrease in precision of the assays. Increases in the temperature of the magnet changed the intensity of the magnetic field and caused the instrument to go out of focus. This occurred with our old LKB 9000, in particular. We found it necessary to warm up the instrument for a couple of hours in the morning before starting the mass fragmentographic work. Obviously the same phenomenon also occurs to some extent with the LKB 2091, at least when higher masses are being monitored. Another factor that greatly influenced the precision and that we noted on all instruments, but studied in some detail with the Varian MAT 112, was the calibration of the switches used for varying the amplification of the channels. This was frequently poor. (The Varian company now pays more attention to this.) It is possible to achieve rather good precision with all four instruments (after some warming up) if the conditions of operation are kept extremely constant the whole time (and especially with at least 5-10 ng amounts of steroid). However, in practice the amounts of biological material being measured change continuously, and different amplifications of the channels have to be used.

APPLICATION OF MASS FRAGMENTOGRAPHY TO METABOLIC STUDIES IN MAN

Table IV

Applications of mass fragmentography to metabolic studies of steroids

1. Absorption of megestrol acetate[12-14]
2. Absorption of estriol[13-15]
3. Estrogens in pregnancy plasma, bile and feces[13,16-19]
4. Effect of ampicillin on estrogen metabolism in pregnancy[20]
5. Intestinal metabolism of 16α-hydroxyestrone[20]
6. Estrogens in non-pregnancy urine and bile[13,16,18]
7. Portal blood androgens[21]

Table IV gives a list of metabolic studies on steroid hormones in man carried out in this laboratory in which selected-ion monitoring was used. Some of these investigations will be described in brief and discussed below. With regard to the last two applications shown in the table the reader is referred to the original publications[13,16,18,21].

Plasma levels of megestrol acetate after oral administration - Comparisons of mass fragmentographic and radioimmunological assays

Megestrol acetate (MA) (3,20-dioxo-6-methyl-4,6-pregnadien-17-yl acetate) is used as an oral or long-acting contraceptive in various devices for the slow release of this steroid. A mass fragmentographic method was developed and used for studies of the plasma levels of the compound after oral administration[12-14]. The compound was chromatographed as the mono-methoxime derivative, and we used medroxyprogesterone acetate (MPA) (3,20-dioxo-6-methyl-4-pregnen-17-yl acetate) as internal standard. However, the retention times of these compounds differ by very little and the problem, discussed above, of interference of the analytical peak signal on one channel with the internal standard peak signal on the second channel sometimes occurred. No antibody to MA is available, but we found that antibodies to MPA could be used, and a radioimmunoassay (RIA) for MA was developed.

Previous RIA studies of plasma MPA have indicated that such systems frequently also measure MPA metabolites[22,23]. Lately more specific methods have been published[14,24]. It was therefore of interest to investigate whether RIA and MF measurements of MA give similar results. After oral administration of 50 mg of MA the plasma level of the compound was measured with both techniques. In Fig. 1 the results of one such experiment are shown. In the early stages of absorption both methods gave results that did not differ significantly from each other, but about 6 h after administration of MA, the RIA gave higher results. Chromatographic experiments indicated that this was due to the presence of metabolites in the plasma[14]. Thus, we may conclude that the specificity of an RIA may be excellent at the beginning of a metabolic experiment but poor at the end and that it may be difficult to demonstrate the non-specificity without comparisons with mass fragmentography. The literature on MPA assays by radioimmunological methods shows that different extraction techniques have resulted in up to 5- to 10-fold differences in results[24].

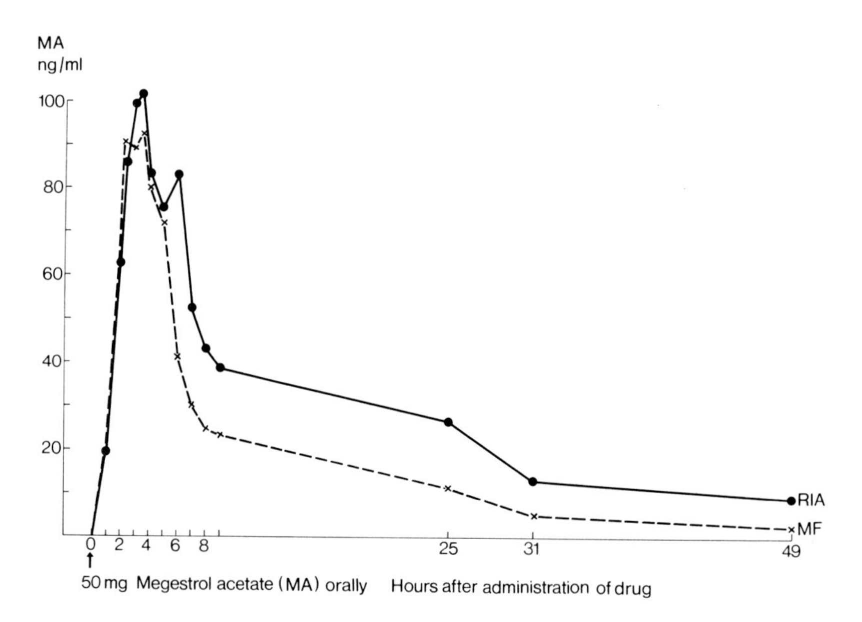

Fig. 1. Megestrol acetate in plasma after oral administration of 50 mg of the steroid. RIA, radioimmunoassay; MF, mass fragmentography.

Even if an RIA gives low values, it does not mean that it is accurate and specific. When comparing RIA and MF of plasma estradiol in pregnancy we observed that the MF values were higher than the RIA values and this discrepancy diminished to some extent with every additional purification step included in the assay before the final RIA. Accurate and specific radioimmunological measurement of estradiol in plasma could only be achieved after chromatographic purification.

Intestinal absorption of orally administered estriol

Several mass fragmentographic methods for the measurement of plasma estriol have been developed and investigated[17] and later used for studies of the plasma level of this estrogen after oral administration of single doses[13,15]. Our main observations were that the level of unconjugated estriol decreased very rapidly after the initial peaking and reached a level of about 0.5-2 µg/l independently of the original dose[15]. This persisted for at least 24 h. The level of conjugated estriol decreased very slowly, and sometimes two peaks were observed. Typical curves have been published previously[13].

Table V

Total amount of various estrogens in maternal plasma at term (mean of duplicate analyses of a pooled sample)[16], maternal bile (one sample, 32nd week of gestation)[26], maternal feces (mean of two 24-h pools from two subjects)[19], in 13 pregnancy urine samples between the 36th and 40th weeks of gestation[27], in bile of post-menopausal women (mean of 3 subjects)[16] and in bile of a 66-year-old man (mean of two samples)[13]

	Pregnancy plasma (term) μg/l	Pregnancy bile (32nd week) μg/l	Pregnancy feces (33-37th week) μg/24h	Pregnancy urine (36-40th week)⁺ μg/24h	Post-menopausal women, bile μg/l	Male bile μg/l
Estriol (E_3)	113	4520	359	25200	1.76	5.64
Estrone (E_1)	52.5	437	97.8	670	0.84	0.12
2-MethoxyE_1	0.6	102	0.5	80	0.39	0.04
Estradiol-17β (E_2)	14.4	42	203	170	0.39	0.67
Estradiol-17α	0.4	22	4.8	30	0.04	n.m.⁺⁺
16-EpiE_3	3.8	115	143	1060	0.75	1.89
17-EpiE_3	0.4	43	38.3	170	0.06	0.05
16α-HydroxyE_1	50.3	3600	3.6	2460	9.24	7.21
16β-HydroxyE_1	8.2	479	1.5	620	1.81	0.45
16-OxoE_2	26.9	307	9.7	1730	1.21	2.21
15α-HydroxyE_1	4.2	179	30.0	220	0.57	0.31
Total	274.7	9846	891.2	32410	17.06	18.59

⁺Analyses were carried out by gas chromatography because of the high amounts of steroids.

⁺⁺n.m., not measured.

Assay of estrogens in plasma, bile and feces during pregnancy

The method used in these studies is a further development of a gas chromatographic procedure[25]. It is possible to measure 11-12 estrogens in various biological fluids of both pregnant and non-pregnant individuals. Values obtained in the last trimester of pregnancy for plasma, bile and urine are given in Table V. For comparison, some biliary levels in post-menopausal women and in one man are also included. The results have been discussed in the original publications[13,16-29].

Effect of ampicillin on metabolism of estrogen in pregnancy

Administration of ampicillin decreased the urinary excretion of estriol and other estrogens in pregnancy[28,29]. The bulk of the decrease of estriol can be accounted for by the decrease in estriol-3-glucuronide excretion[30], a specific intestinal metabolite of estriol[31,32]. On the basis of this finding it seemed likely that the effect of ampicillin on estrogen metabolism is due to an inhibitory effect on the hydrolysis of biliary

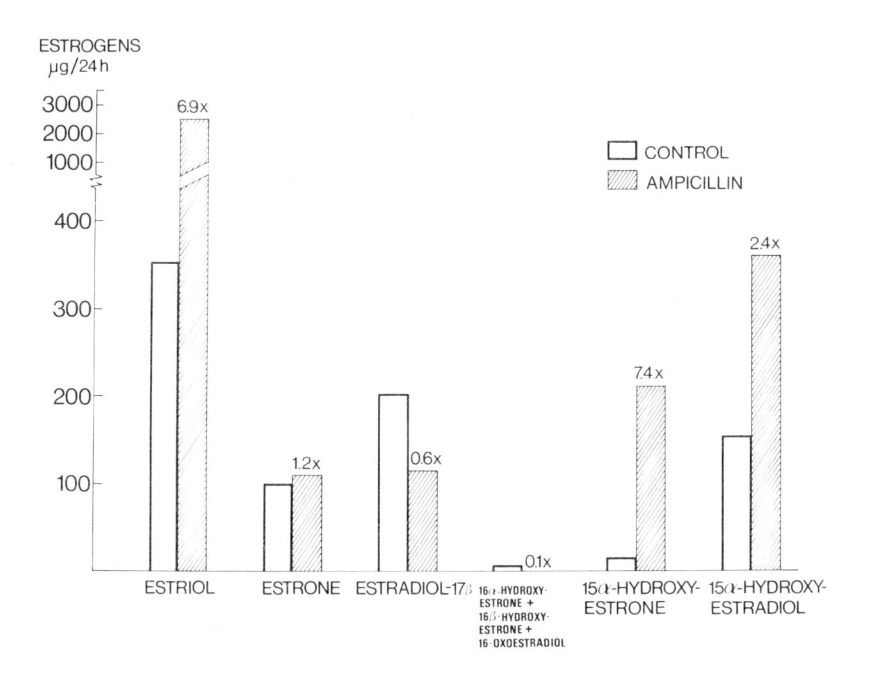

Fig. 2. Effect of ampicillin on fecal unconjugated estrogens in pregnancy. The factor by which the individual estrogens changed after ampicillin is indicated on the top of each ampicillin bar.

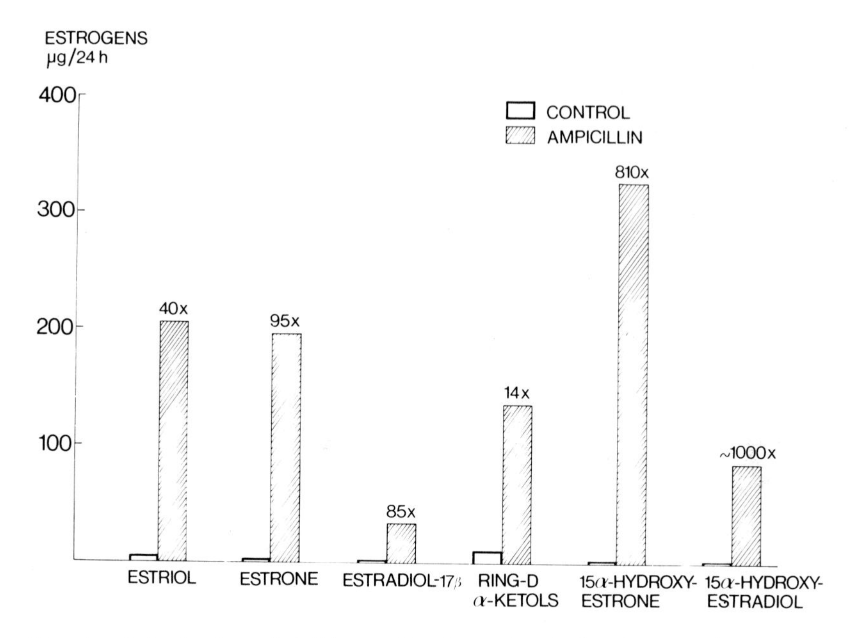

Fig. 3. Effect of ampicillin on fecal conjugated estrogens in pregnancy. The factor by which the individual estrogens changed after ampicillin is indicated on the top of each ampicillin bar.

conjugates through an effect on the intestinal microflora. To clarify this it was necessary to develop a method for determination of estrogen in pregnancy feces. Although the concentrations were rather high, the huge amounts of neutral steroids interfered considerably and precluded the use of ordinary gas chromatography. To achieve specificity, we therefore used mass fragmentography. The results of measurements of fecal estrogens (mean values from two pregnant women) before and after ampicillin are shown in Figs. 2 and 3. The majority of the estrogens in feces were unconjugated. The amounts of estradiol, estrone and 16-epiestriol excreted, relative to the principal estrogen estriol, were greater than in bile or urine in pregnancy, and the ring D α-ketolic estrogens were present in trace amounts only. Considerable quantities of 15α-hydroxyestradiol were also found. Administration of ampicillin caused a huge increase in the fecal excretion of conjugated estrogens. In particular, it caused very striking increases in the excretion of both unconjugated and conjugated estriol, 15α-hydroxyestrone and 15α-hydroxyestradiol. Earlier measurements of the two first-mentioned estrogens in pregnancy urine had shown a marked decrease in connection with administration of ampicillin[29].

These results indicate that ampicillin exerts its effect through inhibition of the hydrolysis of the biliary estrogen conjugates probably by affecting the intestinal microflora.

Fig. 4 shows a comparison between total estrogens in pregnancy bile and in pregnancy feces before and after administration of ampicillin. On the 2nd day of administration many estrogens in feces attained biliary levels. The relative amounts of estrone and especially of estradiol in feces were very high as was the amount of 15α-hydroxyestrone, but those of the labile ring D α-ketolic estrogens were very low. We have therefore postulated that some 15α-hydroxylated estrogens and perhaps estradiol may be formed in the intestinal tract through bacterial action or in the intestinal mucosal cells [20], perhaps from these labile estrogens. Indeed we demonstrated a formation of 15α-hydroxyestrone from 16α-hydroxyestrone by measuring the portal and peripheral plasma estrogens by mass fragmentography after oral administration of 16α-hydroxyestrone (Tables VI and VII)[20]. Some other metabolites were also observed. On the basis of these results we suggest that the following enzymic activities (in respect of estrogens) take place in the intestinal tract: deconjugation, 16-dehydroxylation, 15-hydroxylation, 17β-reduction, 16-oxidation and 16-epimerization.

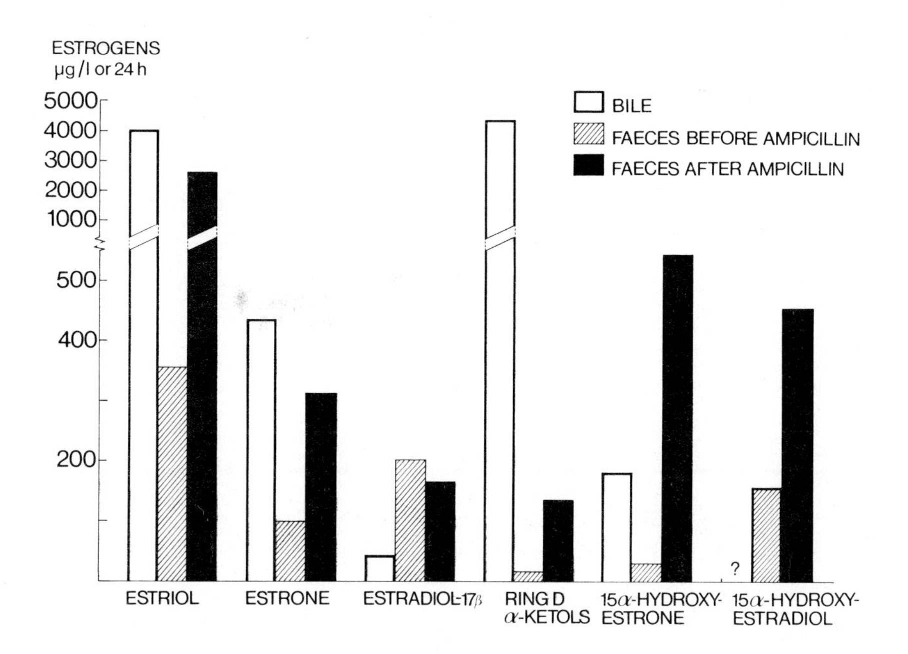

Fig. 4. Comparison between total estrogens in pregnancy bile and in pregnancy feces before and on the 2nd day of ampicillin administration.

On the basis of the experience gained in the above-mentioned investigations it may be concluded that mass fragmentography is a very valuable tool in certain kinds of steroid hormone research. However, at such low levels as are present during metabolic studies of estrogens, the technique is difficult and a lot of experience is required to produce satisfactory results.

Table VI

Estrogens in portal blood after oral administration of 16α-hydroxyestrone (5.4 and 20 mg, respectively) to two post-menopausal women

The values are the highest observed increases in concentration. For meanings of E_1, E_2 and E_3 see Table V.

	Subject 1 (5.4 mg) µg/l	Subject 2 (20 mg) µg/l
Unconjugated E_3	<0.1	<0.1
Conjugated E_3	<0.1	1.5
Unconjugated 16α-hydroxyE_1	<0.1	0.5
Conjugated 16α-hydroxyE_1	4.6	6.3
Unconjugated 16-oxoE_2	8.1	0.2
Conjugated 16-oxoE_2	<0.1	0.3
Unconjugated 15α-hydroxyE_1	16.5	3.4
Conjugated 15α-hydroxyE_1	0.2	0.8
Unconjugated 16-epiE_3	1.5	2.7
Conjugated 16-epiE_3	2.4	0.1

Table VII

Estrogens in peripheral blood after oral administration of 16α-hydroxy-estrone (5.4 and 20 mg, respectively) to two post-menopausal women

The values are the highest observed increases in concentration

	Subject 1 (5.4 mg) μg/l	Subject 2 (20 mg) μg/l
Unconjugated E_3	<0.1	16.0
Conjugated E_3	0.3	1.3
Unconjugated 16α-hydroxyE_1	4.2	1.7
Conjugated 16α-hydroxyE_1	0.3	5.3
Unconjugated 16-oxoE_2	4.3	0.2
Conjugated 16-oxoE_2	0.2	0.8
Unconjugated 15α-hydroxyE_1	0.6	0.5
Conjugated 15α-hydroxyE_1	<0.1	<0.1
Unconjugated 16-epiE_3	<0.1	<0.1
Conjugated 16-epiE_3	<0.1	<0.1

ACKNOWLEDGEMENT

This work was supported by the Ford Foundation, New York.

REFERENCES

1 C.C. Sweeley, W.H. Elliott, I. Fries and R. Ryhage, *Anal. Chem.*, 38 (1966) 1549.
2 H. Adlercreutz, in K. Schubert (Editor), *Abh. Deut. Akad. Wiss. Berlin, Kl. Med.*, 2 (1969) 121.
3 L. Siekmann, H.-O. Hoppen and H. Breuer, *Z. Anal. Chem.*, 252 (1970) 294.
4 H. Breuer, L. Nocke and L. Siekmann, *Z. Klin. Chem. Klin. Biochem.*, 8 (1970) 329.
5 R. Reimendal and J. Sjövall, in V.H.T. James and L. Martini (Editors), *Proceedings of IIIrd International Congress on Hormonal Steroids, Hamburg, 1970*, Excerpta Medica, Amsterdam, 1971, p. 228.
6 C.J.W. Brooks and B.S. Middleditch, *Clin. Chim. Acta*, 34 (1971) 145.
7 R.W. Kelly, *J. Chromatogr.*, 54 (1971) 345.
8 L. Palmér and B. Holmstedt, *Science Tools, The LKB Instrument Journal*, 22 (1975) 25.
9 F.C. Falkner, B.J. Sweetman and J.T. Watson, *Appl. Spectrosc. Rev.*, 10 (1975) 51.
10 H. Adlercreutz and D.H. Hunneman, *J. Steroid Biochem.*, 4 (1973) 233.
11 M.G. Lee and B.J. Millard, *Biomed. Mass Spectrom.*, 2 (1975) 78.
12 H. Adlercreutz, U. Nieminen and H.-S. Ervast, *J. Steroid Biochem.*, 5 (1974) 619.

13 H. Adlercreutz, F. Martin, O. Wahlroos and E. Soini, *J. Steroid Biochem.*, 6 (1975) 247.
14 F. Martin and H. Adlercreutz, in S. Garattini (Editor), *Symposium on the Pharmacology of Steroid Contraceptive Drugs, Milan, 1976.* Raven Press, New York, in press.
15 J. Wähäpassi and H. Adlercreutz, *Contraception*, 11 (1975) 427.
16 H. Adlercreutz, M.J. Tikkanen and D.H. Hunneman, *J. Steroid Biochem.*, 5 (1974) 211.
17 H. Adlercreutz and P. Nylander, *Biomed. Mass Spectrom.*, 1 (1974) 332.
18 H. Adlercreutz, in A. Frigerio and N. Castagnoli (Editors), *Mass Spectrometry in Biochemistry and Medicine*, Raven Press, New York, 1974, p. 165.
19 H. Adlercreutz and F. Martin, *Acta Endocr.*, (1976) in press.
20 H. Adlercreutz, F. Martin, M. Pulkkinen, H. Dencker, U. Rimér, N.-O. Sjöberg and M.J. Tikkanen, *J. Clin. Endocrinol.*, (1976) in press.
21 F. Martin, H. Adlercreutz, B. Lindström, H. Dencker, U. Rimér and N.-O. Sjöberg, *J. Steroid Biochem.*, 6 (1975) 1371.
22 J.C. Cornette, K.T. Kirton and G.W. Duncan, *J. Clin. Endocrinol.*, 33 (1971) 459.
23 M.E. Royer, H. Ko, J.A. Campbell, H.C. Murray, J.S. Evans and D.G. Kaiser, *Steroids*, 23 (1974) 713.
24 M. Hiroi, F.Z. Stanczyk, U. Geobelsmann, P.F. Brenner, M.E. Lumkin and D.R. Mishell, Jr., *Steroids*, 26 (1975) 373.
25 H. Adlercreutz, in H. Breuer, D. Hamel and H.L. Krüskemper (Editors), *Methods on Hormone Analysis*, Georg Thieme Verlag, Stuttgart, and John Wiley & Sons, New York, Sydney, Toronto, 1975, p. 480.
26 H. Adlercreutz, H.-S. Ervast, A. Tenhunen and M.J. Tikkanen, *Acta Endocr.*, 73 (1973) 543.
27 H. Adlercreutz and T. Luukkainen, *Acta Clin. Res.*, 2 (1970) 365.
28 K. Willman and M.O. Pulkkinen, *Amer. J. Obstet. Gynecol.*, 109 (1971) 893.
29 H. Adlercreutz, F. Martin, M.J. Tikkanen and M. Pulkkinen, *Acta Endocr.*, 80 (1975) 551.
30 M.J. Tikkanen, M.O. Pulkkinen and H. Adlercreutz, *J. Steroid Biochem.*, 4 (1973) 439.
31 K. Dahm and H. Breuer, *Z. Klin. Chem. Klin. Biochem.*, 4 (1966) 153.
32 K.F. Støa and M. Levitz, *Acta Endocr.*, 57 (1968) 657.

TECHNIQUES FOR QUANTITATIVE MEASUREMENTS BY MASS SPECTROMETRY

C.C. SWEELEY, S.C. GATES, R.H. THOMPSON, J. HARTEN, N. DENDRAMIS and J.F. HOLLAND

Department of Biochemistry, Michigan State University, East Lansing, Mich. (U.S.A.)

The basic principles of mass spectroscopy have been known since early in the present century. The initial qualitative applications were uniformly successful, largely because the accuracy of the mass axis measurement was surprisingly good. Great effort was therefore directed to instrumental refinements in mass measurement, resulting in the relatively rapid development of excellent low- and high-resolution mass spectrometers. However, early applications of the instrument for quantitative analyses were greatly encumbered by problems of sample injection, ion detection and extremely long analysis times.

The development of electronic theory produced phototubes, which enabled reasonably accurate photoplate measurements, and Faraday cups for accurate ion beam measurements. The electrical nature of the Faraday cup permitted the direct recording in real-time of the ion beam response curve that is obtained during mass scanning. The subsequent development of electron multipliers furnished a sensitivity increase of about a million-fold and permitted analyses to be made by mass spectrometry with trace amounts of sample.

Investigators continued to utilize the improvements of electronic and vacuum technology in their instruments; however, for the most part the mass spectrometer remained a physicist's "gadget", with remarkably few analytical applications up to the late 1950's. This is an amazing situation when one considers its inherent potential. In 1960, the mass spectrometer was successfully connected to a gas chromatograph[1] and its role subjugated to that of an elution detector. This proved to be an excellent marriage of diverse instrumental capabilities - the mass spectrometer brought sensitivity and another dimension (mass) to the analysis of organic compounds while the gas chromatograph provided a convenient sample injection capability for the mass spectrometer, with the added benefit of a time-based separation of the components of impure samples as well as complex mixtures. The combined instrument (GC-MS) became a remarkable analytical tool and was rapidly assimilated for chemical analyses in nearly every discipline of science. It soom became so prolific, in fact, that the growth in applications was limited only by the time required to make use of the copious amounts of data that were produced in short time intervals.

The obvious next step in the development of this instrumentation was its combination with the solid state computer. These devices, both big and small, furnished the capability for collection, storage and processing of the data from GC-MS analyses and allowed the analyst to begin to contemplate use of the full analytical potential of the mass spectrometer. For qualitative analysis, for example, the ion response curve could now be studied in both the time and mass coordinates simultaneously, with the subsequent generation of two-dimensional reference schemes using mass and retention time as complementary criteria in the identification procedures, as will be discussed in more detail below. Factors which affect the quantitation parameters of accuracy, precision, range and sensitivity are more complex; however, it can be stated without reservation that the interface between the GC-MS instrument and the computer for on-line data collection and processing has created one of the most exciting and useful measurement systems of modern science.

For convenience, the analytical factors may be divided into three classifications: sampling factors, instrumental factors and statistical factors. The effects of these factors on quantitation can be illustrated by specific examples, using as extreme cases the optimization of versatility (cyclic scanning) and the optimization of sensitivity and accuracy (selected ion monitoring). The first method utilizes rapid repetitive scanning of the mass spectrometer over a large mass range throughout the time of the gas chromatographic run, the measurement time being divided proportionately throughout the mass axis during each scan. This technique is statistically inefficient for the quantitation of ion intensity since much of the time is spent measuring masses of no particular pertinence to the analysis as well as time spent in regions of the scan axis between masses. It does accommodate complex samples, however, as all eluting peaks will have a representative mass spectrum recorded and the presence of a compound need not be anticipated prior to its elution from the gas

TYPICAL MSSMET LIBRARY INPUT

```
*HOMOVANILLIC
1925
326,1
209,308,311,152,326,280
```

Fig. 1. MSSMET library input for trimethylsilyl (TMS) derivative of homovanillic acid. Entry includes name, retention index, designate ion, K-factor and set of confirming ions paired with their relative intensities.

chromatographic column. Data from these multiple scans are usually stored in the computer system for processing after the run has been completed. Outputting in the form of plots of selected ion intensities *versus* scan number is a method that Hites and Biemann referred to as mass chromatography[2]. It is possible also to produce a limited number of mass chromatograms in real-time with some data systems[3].

The technique of mass chromatography has been exploited by a computer program (MSSMET) in development in our laboratory for fully automated qualitative and quantitative analysis of multi-component mixtures of metabolites from biological fluids[4-6]. The program employs a reverse library search routine wherein each compound of interest is characterized by a retention index value and the *m/e* value of a specific ion, called the designate ion, that distinguishes the compound from other substances in the mixture with approximately the same chromatographic properties. The presence of additional characteristic ions confirms the identification of the compound in the sample. Peak height and area calculations of the designate ion are compared with those of an internal standard to give relative amounts of the compound, and dimensional factors are applied to obtain absolute quantitation of the substance.

Early versions of MSSMET utilized a PDP-8/I computer (Digital Equipment Company); however, a more versatile version has been developed with a PDP-11/40 computer (Digital), where most of the program has been written in standard Fortran IV[7]. This latter version makes possible the completely computerized peak identification and quantitation. Fig. 1 illustrates the type of information entered into the computer to establish a library reference for a particular compound. In addition to the name of

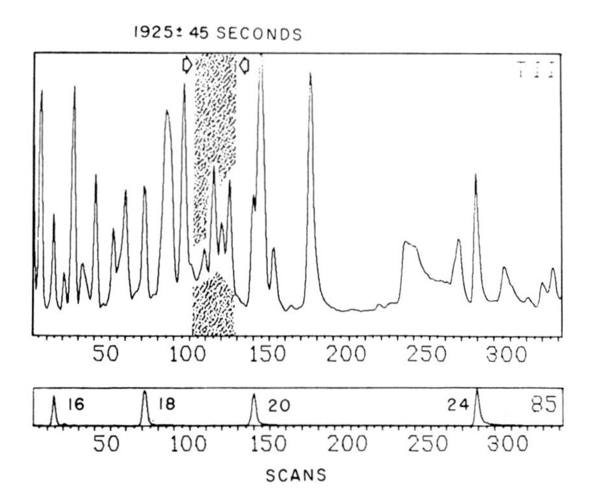

Fig. 2. Location of the retention index "window" for the TMS derivative of homovanillic acid as described in text. Sample is acidic urinary metabolites on 10 ft. 5% OV-17, programmed at 4°/min from 160-280°.

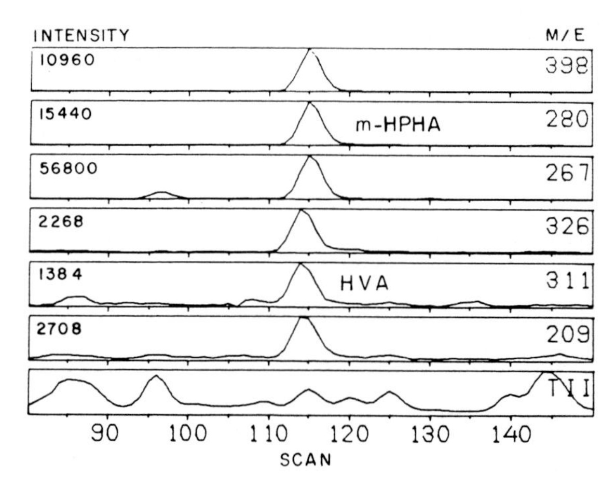

Fig. 3. Mass chromatograms illustrating complete resolution of TMS derivatives of homovanillic acid and *m*-hydroxyphenylhydracrylic acid. Scans shown are from sample in Fig. 2.

TYPICAL MSSMET OUTPUT

```
HOMOVANILLIC
  2 +  96    8035.  0.949E-01 12:28   0:00 1924   -1  109  114  123
    +  98    2209.  0.731E-01
```

Fig. 4. MSSMET output indicating confirmation of presence of homovanillic acid in urine sample shown in Fig. 2. Second line of output includes peak number, overall match category, correlation coefficient, area, relative quantity, retention time, deviation of retention time from expected value, retention index, deviation of retention index from library (expected) value, and scan numbers indicating the beginning, maximum and end of the peak. The third line includes corresponding information based on peak height instead of peak area.

the compound, the designate ion and confirming ion set (as determined from spectra of pure reference compounds), the library entry contains a retention index value and a quantitative factor (K). The retention index is determined by interpolation between hydrocarbon standards coinjected with the sample[8]. The value of K relates the area of the designate ion of the compound to the designate ion of the internal standard and allows corrections for the dimensional factors necessary for quantitation in absolute terms. The location of the "window" or region within which the compound must elute is illustrated in Fig. 2; this process uses the library retention index value as the center point of the window. As shown in Fig. 3, the computer then collects mass chromatograms of the designate and confirming ions within the window, and uses them to determine the presence or absence of the compound. The area of each peak of the designate ion within the window is also calculated for use in quantitation.

To ascertain if a peak of the designate ion actually represents the compound, a correlation coefficient is calculated by comparison of the library value for that compound. The combination of the correlation coefficient and the deviation of the retention index from the library value is used to judge which peak represents the compound of interest. These two criteria are combined into an overall match category of (+), (?) or (-), which is then included as part of the MSSMET output, as illustrated in Fig. 4. The computer takes all peaks in the "+" category and stores them in a permanent file for subsequent statistical comparisons between individual urine samples.

An extremely important factor in assuring reliable identification is the use of precise retention indices. As shown in Fig. 5, these retention indices are reproducible to better than 0.1%, which indicates that the

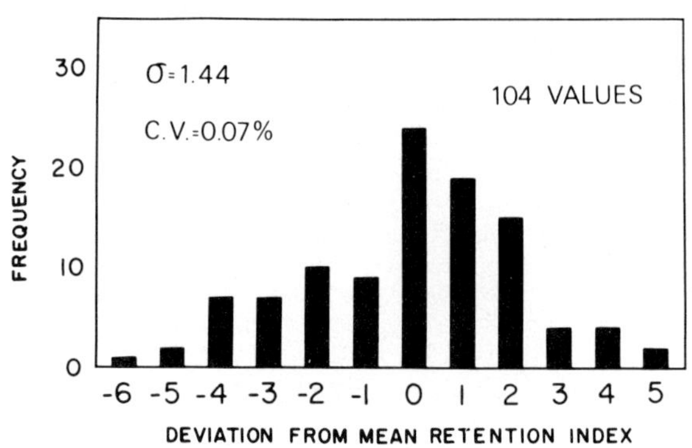

Fig. 5. Precision of MSSMET determination of retention indices based on injections of pure derivatized compounds at 20 ng to 10 µg.

TYPICAL MSSMET LIBRARY INPUT

```
*3A,17A,21-TRIHYDROXY-5B-PREGNANE-11,20-DIONE(THE)
2974
578,1
578,39,488,22,609,10,506,7,398,8
```

Fig. 6. MSSMET library input for methyloxime-TMS derivative of tetrahydrocortisone (THE).

misidentification of a peak is nearly eliminated by this criterion alone, and is even less likely when combined with the correlation coefficient.

Another example of MSSMET is shown in Figs. 6-8. The compound of interest is a steroid, tetrahydrocortisone (THE), excreted in human urine at a level of approximately 1 mg/day. This steroid was identified in a urine sample on the basis of the information that had been entered previously into the library file in the format shown in Fig. 6. The behavior of the confirming ion set in the vicinity of the peak is shown graphically in Fig. 7, which indicates the conformity of the confirming ion profiles with that of the designate ion at *m/e* 578. As indicated in Fig. 8, the program identified and quantitated the methoxime-trimethylsilyl derivative of this metabolite with a correlation coefficient of 95%. Presently, the MSSMET library contains 30 steroid entries that can be automatically analyzed in urine samples. Mass chromatograms of these compounds may be plotted together to facilitate recognition of several

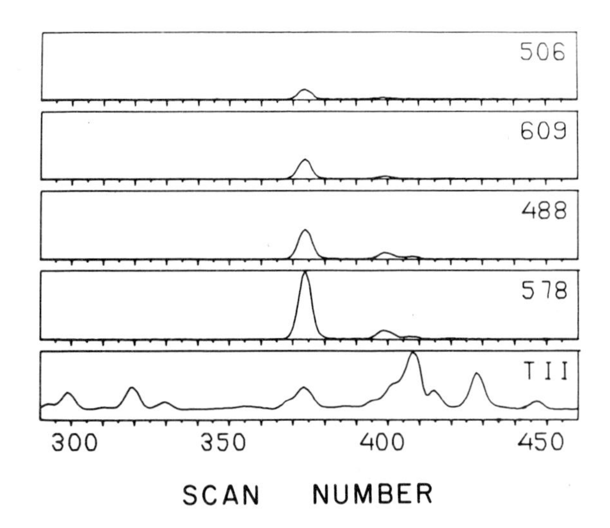

Fig. 7. Mass chromatograms of designate and confirming ions for THE in human urine. Sample was subjected to XAD-2 extraction, enzymatic hydrolysis and methyloxime-TMS derivatization and run on a 12 ft. 1% SE-30 column programmed from 200 to 280° at 2°/min.

TYPICAL MSSMET OUTPUT

```
3A,17A,21-TRIHYDROXY-5B-PREGNANE-11,20-DIONE(THE)
  2 +  95   42912. 0.171E 01 29:51   0:03 2972   -2  370  379  391
    +  94    8235. 0.181E 01
```

Fig. 8. MSSMET library output for sample shown in Fig. 7.

steroid components at the same time as well as to study their relative levels in the urine, as shown in Fig. 9. The range of concentrations for 15 steroids in human urine, from a study involving 11 random normal subjects, is illustrated in Fig. 10. Investigations of this type can be extremely important in pinpointing metabolic abnormalities in diseases of unknown etiology. These studies are greatly facilitated by highly automated programs such as MSSMET and those developed by other investigators[9-11].

Improvements in repetitive scanning

There are several ways in which the quantitative viability of rapid repetitive scanning techniques can be enhanced. High-speed scanning over a more limited mass range reduces the total time of a scan and can increase statistical reliability. This necessarily introduces a

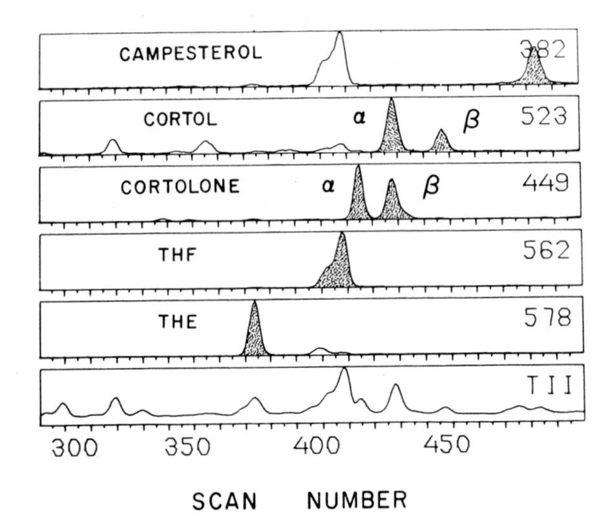

Fig. 9. Mass chromatograms of designate ions of steroids in human urine. The compounds are methyloxime-TMS derivatives of tetrahydrocortisone (THE), tetrahydrocortisol (THF), cortolone, cortol and campesterol (quantitative standard).

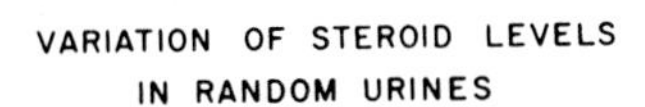

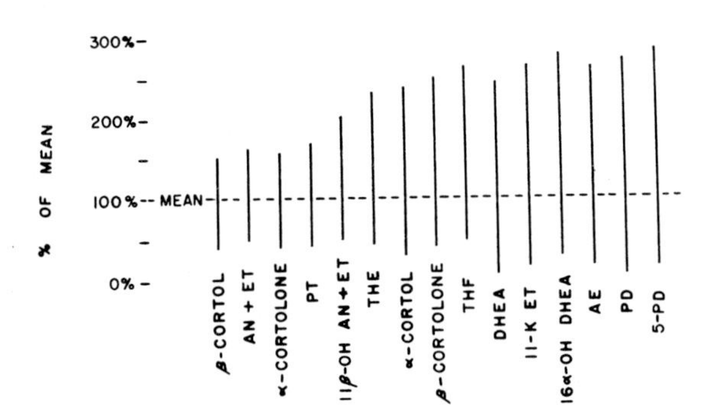

Fig. 10. Variation of steroid levels in 11 human urine samples. Values were calculated by MSSMET, normalized to creatinine levels, and expressed as percent of the normalized mean.

limitation in versatility, which can be minimized, however, by altering the mass range scanned as a function of elution parameters. For example, during the early part of an elution, the high mass range can be omitted since large molecules would not be expected to have short elution times. Later, when the GC peaks are broader, the scans can be extended progressively to higher masses without sacrificing accuracy. In other situations, threshold sensitivities may be employed to reduce the number of ions detected per scan, greatly assisting data processing. Post-run processing time may be lengthened to permit the use of sophisticated data smoothing routines and mathematical fitting functions that may engender higher levels of accuracy. Summing of several ions originating from the same sample molecule can be used to increase sensitivity to levels approaching that of SIM. In addition GC parameters may be altered to produce broader, easier-to-quantitate peaks. An obvious factor in each of these approaches is the trade-off between qualitative and quantitative needs; the ability to analyze complex samples continually works against the limits of sensitivity and accuracy.

Selected ion monitoring

At the opposite extreme in quantitative analysis by mass spectrometry are those methods where the mass spectrometer spends more and more of the available measurement time on fewer and fewer ions. The most favorable statistical situation occurs when all of the time is spent on measurements of the intensity of a single ion (single ion monitoring). This technique was first described by Henneberg in 1961 (ref. 12) and represents the first attempt at mass-specific GC detection. Time-dependent variations in the performance of the mass spectrometer, such as sample mobility, ionization efficiency, vacuum fluctuations and electronic stability, introduce errors that often make this method of analysis undesirable. For specific applications requiring the maximum possible accuracy, such as certain isotope ratio determinations, the problem has been resolved using dual detectors to simultaneously record the intensities of two ions, the ratio of which will cancel out instrumental variations. These systems are not particularly versatile, however, and as a result have received little attention in GC-MS applications.

Instead, instrumental methods have been developed that utilize one detector with the focus of the mass spectrometer being alternated between two or more *m/e* values in a timed sequence. With magnetic sector mass spectrometers the focusing changes may be effected by switching either the magnetic field or the accelerating voltage; early applications of this approach with an accelerating voltage alternating (AVA) device made possible stable isotope analyses in GC-MS as well as analyses of partially resolved compounds[13] and led to the truly prodigious application of selected ion monitoring (SIM) in pharmacology, biochemistry and other fields as well illustrated by the presentations at this symposium. Selected ion monitoring has also been accomplished with quadrupole mass spectrometers by switching the mass filter electrostatic voltages.

A unique statistical problem arises from the time-sharing of the detector serially by two or more ions. This problem results because there are actually two quantities that are being measured, ion intensity (direct measurement) and GC elution profile (indirect measurement). The detector provides a direct measure in real time of the ions impinging upon it. Since these ions are formed, focused, accelerated and separated in a continuous sequence, their appearance at the detector will exhibit the combined normal statistical variations inherent in these processes as well as the statistical variation of intensity that results from the partition process in the GC column. To obtain an accurate measurement of the intensity of a specific *m/e* value, it is desirable to remain in focus on that ion for a sufficient time duration. The optimum duration will vary with the rate at which the ions are colliding with the detector (ion current), the smaller the ion current, the longer the optimum duration. Most of the analytical information resides in the GC elution response curve, however, which must be constructed for each ion of interest by creating a function using the time-dependent individual ion intensity measurements for that specific ion. The accuracy with which retention time, peak height and peak area can be measured all depend on the number of data points defining the GC peak. Unfortunately, these two considerations are in opposition - the more rapidly focusing is switched between the various ions the more points in the reconstructed GC peak, but with reduced accuracy of each individual measurement of ion intensity. This is a situation which can be significant when weak ion beams are encountered. Conversely, longer and therefore more accurate ion measurements result in less points defining the shape of the GC peak and hence less accurate peak heights and areas. An additional instrumental factor is encountered in the time required after each transition to allow the ion optical system to "settle in" on the next ion. This time is lost to the measurement and introduces a bias against high-frequency switching. It is apparent that the optimum switching frequency will vary from instrument to instrument

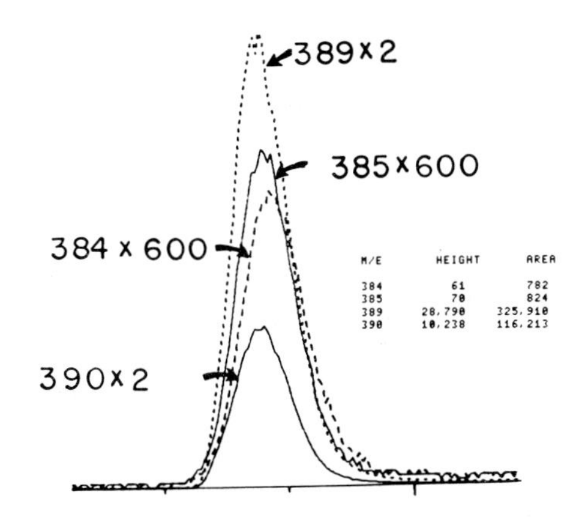

Fig. 11. Selected ion monitoring traces of the TMS derivative of 3,4-dihydroxyphenylacetic acid, with 1.2 ng of protium form and 1.0 μg of a pentadeuterio form injected. The data for peak heights and areas are uncorrected for the blanks.

and from application to application, with the result that this parameter must always be considered when developing analytical procedures based on SIM analyses.

In a typical operation of SIM with a magnetic sector mass spectrometer, the magnet is set to a specific value and kept constant. The ions selected for measurement are sequentially switched into focus by changes in the accelerating voltage. In our system the changes are performed under the direction of a mini-computer and one to eight ions may be chosen for a single SIM analysis. An additional feature of this system permits the computer to maintain a fine focus on each of the ion beams by a routine which utilizes small voltage increments of the accelerating voltage with closed-loop intensity measurement feedback to continually insure optimum focus[14]. The analytical excellence of this technique is illustrated in Fig. 11, in which ion intensities in the ng and μg range are simultaneously and accurately measured. These results are typical of the dynamic range and accuracy obtained with SIM.

Despite its analytical potential, SIM has suffered from several disadvantages which have limited its applicability, even in situations

Table I

Disadvantages of SIM for quantitation of organic compounds by GC-MS

- limited number of compounds that can be assayed per GC run
- difficulties in verifying peak identity, especially at low levels
- need for labeled internal standards
- inability to demonstrate the presence of "unexpected" compounds
- limited useful mass range with AVA devices on magnetic sector instruments

where only quantitative information is desired. These disadvantages, which are summarized in Table I, are particularly acute when dealing with biological samples that contain a large number of peaks, many of which may be of interest. However, most currently available GC-MS-COM systems are limited to monitoring one set of approximately eight ions per sample, a situation which precludes the quantitation of more than a very small fraction of the compounds in the sample. In addition, it is often difficult to select even a single ion that is unique for the compound of interest. Hence, especially at low concentrations of the compound, the identity of the peak being quantitated is often uncertain. Other complications are the lack of deuterated internal standards for many of the biologically interesting compounds and the fact that SIM generally is unable to indicate the presence of metabolites other than those specifically anticipated by the researcher. A final disadvantage is inherent in SIM using magnetic sector instruments. Large accelerating voltage changes produce ion source defocusing, resulting in a loss of sensitivity which, in effect, reduces the useable mass range of the instrument.

Improvement of SIM

Many of the disadvantages of SIM discussed above can be eliminated without appreciable loss of quantitative veracity. For example, the limit in the number of compounds that can be assayed in a single GC run can be increased by a technique which we have named multiple ion set-selected ion monitoring (MIS-SIM). In magnetic sector instruments this has been accomplished by an alteration of the magnetic field setting several times during the course of a single GC run[15]. At each magnetic field setting, SIM analysis is performed on a preselected ion set using accelerating voltage focusing. Since the maximum time required to establish a new setting of the magnet is about 8 seconds, peaks as close as one minute apart in retention time may be quantitated with ease. An example of this

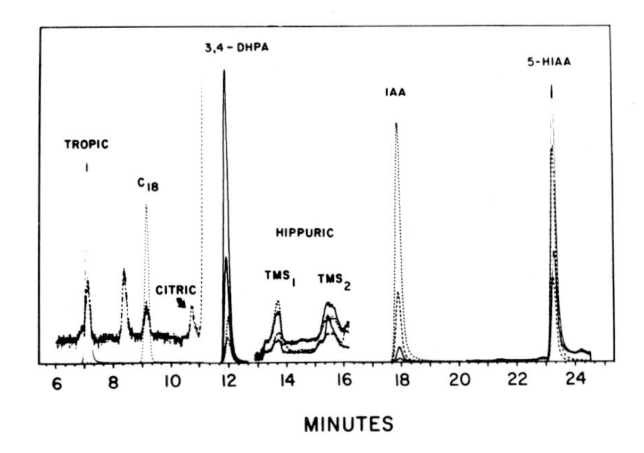

Fig. 12. Typical computer plot of data from multiple ion set-selected ion monitoring (MIS-SIM) using a magnetic sector mass spectrometer (LKB 9000). Compounds were run as TMS derivatives on 10 ft. 5% OV-17 programmed 160-280° at 4°/min. Compounds include tropic acid, *n*-octadecane, citric, 3,4-dihydroxyphenylacetic, hippuric, indoleacetic and 5-hydroxyindoleacetic acids.

technique is shown in Fig. 12, where seven compounds are quantitated during a single GC elution, using five different magnet settings.

A substantial increase in the qualitative ability of MIS-SIM can be realized by using a program called RI-SIM to identify eluting components by use of retention indices in a small reference library in the computer[16]. This retention index-based selected ion monitoring technique, illustrated in Fig. 13, depends upon the detection of two or more early eluting components of the mixture whose retention indices are known. From the retention times observed and the indices of these compounds the computer calculates, by extrapolation, the retention time of each succeeding peak from library retention indices. The observed occurrence of these peaks may then be used during the run to provide a retention time *versus* retention index curve that may be used for succeeding calculations. Using the expected retention time of the next peak, the program can predict the exact time to switch to the next appropriate magnet field setting and voltages for the selected ions in this set, and also can predict the location of the beginning and apex of each GC peak of interest. These data allow the operator to force the computer to initiate its focusing algorithms, which effectively coordinates automatic focusing with the occurrence of the peak.

As shown in Table II, these extrapolated retention times are extremely precise, and are particularly useful in identifying small peaks. However,

RISIM OUTPUT

```
   86.4     1861.       31.3       22.5       -4.3       -3.1
   89.3     1863.       28.4       20.5       -7.1       -5.1
   92.1     1865.       25.6       18.4       -9.9       -7.2
   94.9     1867.       22.8       16.4      -12.8       -9.2
   97.7     1869.       20.0       14.4      -15.6      -11.2
WITHIN THE NEXT  41 SECONDS, TYPE EITHER A 1 TO
INCLUDE PEAK IN RETENTION INDEX TABLE OR A 0 NOT TO INCLUDE IT>1
  CUR T    /CUR RI    /T NXT PK/RI NXT PK/T SWITCH /RI SWITCH
  122.0     1886.       54.1       38.6       20.4       14.5
  124.8     1888.       51.3       36.6       17.6       12.5
  127.6     1890.       48.5       34.6       14.8       10.5
  130.4     1892.       45.7       32.6       11.9        8.5
  133.3     1894.       42.9       30.6        9.1        6.5
  136.1     1896.       40.1       28.6        6.3        4.5
  138.9     1898.       37.3       26.5        3.5        2.5
  141.7     1900.       34.4       24.5        0.7        0.5
  144.5     1902.       31.6       22.5       -2.1       -1.5
```

Fig. 13. Portion of a typical RI-SIM program output, generated in real time during a MIS-SIM analysis of the series of samples described in Fig. 12. The output includes elapsed time in seconds since the injection, current retention index, estimated arrival of the next peak of interest in both seconds and retention index units, and the time prior to magnetic switching in both seconds and retention index units. Program also queries the operator as each peak appears about whether to update the retention index *versus* time table.

Table II

Precision of retention index (R.I.) determination by RI-SIM[+]

Compound[++]	Mean		Standard deviation		Coefficient of variation(%)	
	R.I.	Time (sec)	R.I.	Time (sec)	R.I.	Time
Citric	1879	708	2.0	16	±0.11	±2.3
3,4-DHPA	1947	786	6.2	20	±0.32	±2.5
IAA	2241	1139	6.3	16	±0.28	±1.4

[+] Based on linear extrapolation for the retention times of tropic acid (R.I. = 1695) and *n*-octadecane (R.I. = 1800) for 10 determinations.

[++] 3,4-DHPA = 3,4-dihydroxyphenylacetic acid; IAA = indoleacetic acid.

it should be noted that if a single linear extrapolation from only two initial points, early in the GC run, is used to predict all of the succeeding retention times, the observed retention indices will not be the same as those determined by linear interpolation between closely spaced hydrocarbons.

Again, it should be noted that efforts to expand the versatility of SIM inevitably place restraints on its analytical capability and the extent of the modification employed will ultimately be determined by analytical need.

Comparison of quantitative methods using SIM and MSSMET

For the first part of these studies, a serial dilution of several organic acids (citric, 3,4-dihydroxyphenylacetic, homovanillic, ascorbic, *p*-hydroxycinnamic, indoleacetic and 5-hydroxyindoleacetic acids) was prepared. This dilution included 15 samples at levels from 1.2 ng to 10.0 μg injected, plus a blank, to all of which an internal standard (tropic acid) was added at a level of 1.00 μg injected. In addition, five compounds were added as deuterated carriers to the same mixtures (3,4-dihydroxyphenyl-d_3-acetic-d_2 acid, *a*-d_2-hippuric acid, *a*-d_2-homovanillic acid, *a*-d_2-5-hydroxyindoleacetic acid, and *a*-d_2-indoleacetic acid from Merck, Sharp and Dohme, Canada) at levels of 1.00 μg injected. These samples were examined by both MIS-SIM and MSSMET; in the former case the RI-SIM program was also used. TMS derivatives were made in all cases using a 1:9 mixture of redistilled dry pyridine and bis-trimethylsilyl-trifluoroacetamide (Pierce).

The first parameter to be investigated involved precision. Clearly, the best precision from either technique should be obtained during the measurement of isotope ratios of relatively intense peaks. For this purpose, the ratio of two peaks in the molecular ion cluster $\{P/(P+1)\}$ or some other intense ion was determined for each of the deuterated compounds measured. As shown in Table III, the precision (coefficient of variation) was better than 0.35% for SIM, while repetitive scanning with MSSMET analysis had a corresponding precision of less than 3.0%. Generally, the agreement of the ratios obtained by each method was quite good. The measurements were made only on samples for which there was little or no contribution from the protium form to the intensities measured for the deuterated sample. Since both ions were from the same molecular species, these results represent an evaluation of instrumental precision and accuracy for the two methods.

The linear range and sensitivity were the next quantitative variables to be examined using the two types of techniques. As shown in Figs. 14-16, the most sensitive and linear technique is the magnetic field switching version of selected ion monitoring (MIS-SIM) with isotope carrier as the internal standard. This technique was linear over the entire range of dilution, using the pentadeuterio form of 3,4-dihydroxyphenylacetic acid, once a correction was made for the blank. Appropriate corrections to obtain linear curves were more complicated for the other deuterated compounds which contained only two deuterium atoms per molecule. In these cases, it was adequate to utilize a non-linear standard curve. This

Table III

Precision of isotope ratio determination[+]

Compound[++]	*m/e*	Mean ratio		Coefficient of variation(%)	
		MSSMET	MIS-SIM	MSSMET	MIS-SIM
3,4-DHPA-d_5	389/390	2.82	2.81	$\pm$2.2	$\pm$0.23
IAA-d_2	321/322	3.25	3.43	$\pm$2.8	$\pm$0.36
5-HIAA-d_2	409/410	2.56	2.56	$\pm$1.4	$\pm$0.33

[+] Based on 10 injections of 1.0 µg of each compound.

[++] 3,4-DHPA-d_5 = labeled 3,4-dihydroxyphenylacetic acid; IAA-d_2 = labeled indoleacetic acid; and 5-HIAA-d_2 = labeled 5-hydroxyindoleacetic acid.

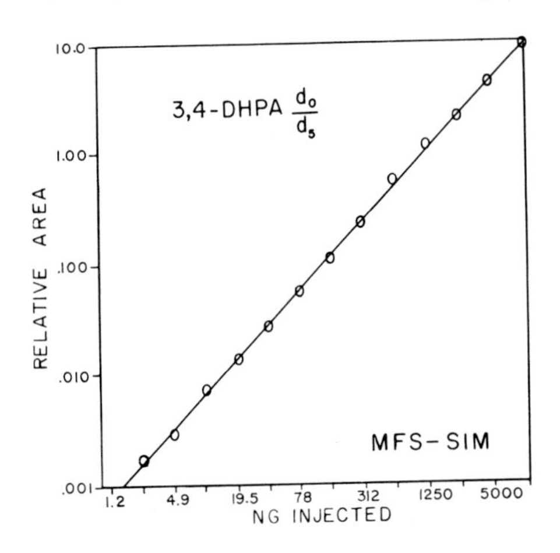

Fig. 14. A quantitative working curve for a series of TMS derivatives of 3,4-dihydroxyphenylacetic acid injected with 1.0 µg of d_5 carrier. The magnetic field switching program was used for this analysis.

assumption has been verified by the linear curves obtained when the values from MSSMET were plotted against those from SIM analyses.

MSSMET-based quantitative analysis was much less sensitive than SIM. Depending on the type of sample, the limit of sensitivity was in the range from 10-20 ng injected, as compared to the sub-nanogram sensitivities of SIM. Isotope dilution techniques did not appear to increase the sensitivity of the MSSMET approach.

The ease of compound identification, a qualitative feature, was also compared. As previously shown[4-6], MSSMET was able to identify peaks with

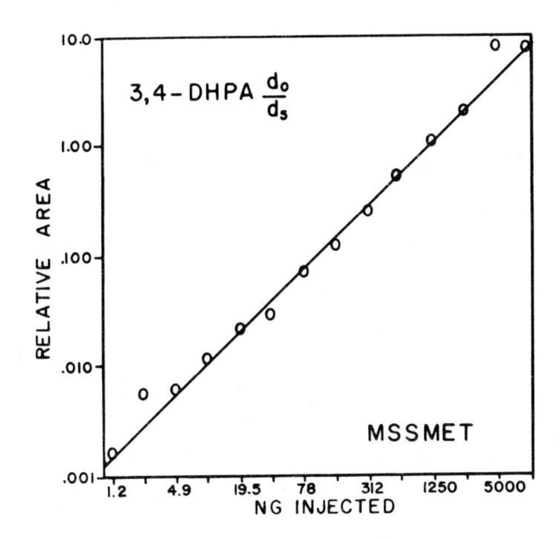

Fig. 15. Quantitative working curve produced by repetitive scanning and MSSMET analysis of the data. The samples analyzed were the same as those used in the study by MIS-SIM, shown in Fig. 14.

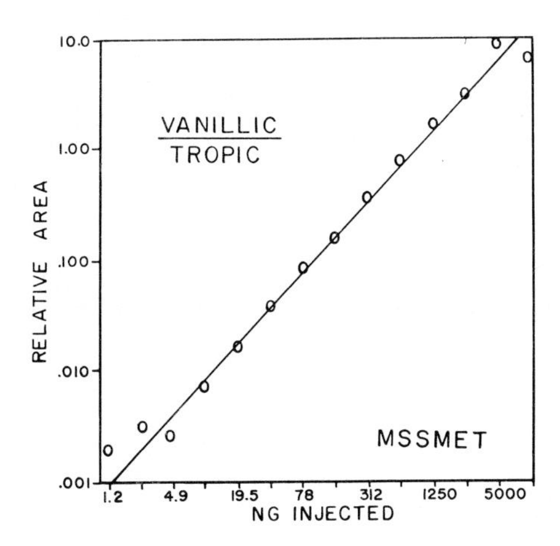

Fig. 16. Quantitative working curve generated by MSSMET analysis of varying amounts of the TMS derivative of vanillic acid; 1.0 µg of a TMS derivative of tropic acid was co-injected and used as an internal standard.

a correlation coefficient greater than 85% down to the limits of its sensitivity. In contrast, SIM techniques have not generally been designed to provide easy compound identification. The RI-SIM program, however, was able to predict the location of each peak of interest in real time, as shown in Fig. 8. It also predicted the optimum time to switch to the next ion set. The location of a peak can be predicted accurately even when extrapolating to distant peaks (Table II). Using tropic acid (R.I. = 1695) and octadecane (R.I. = 1800) as indicators, the indoleacetic acid retention time was predicted to within 6.3 R.I. units (0.25%) while the actual retention varied by 1.4%.

As shown in the summary comparison (Table IV), SIM methods appear to be superior to cyclic scanning for quantitative analysis, but are decidedly less versatile in qualitative applications. Our recent developments on magnetic sector GC-MS-COM interfaces and programs, described briefly in this presentation, have produced an increase in the useable range for selected ion monitoring and have increased the number of compounds that can be analyzed from a single sample with little loss in quantitative reliability. The practical limit to the number of serial SIM analyses that can be obtained from one GC injection is determined basically by the chromatographic separation and knowledge of the constituents of the sample mixture. The resolution, mass range, and sensitivity of magnetic sector instruments coupled with the outstanding potential of SIM have consistently encouraged these developments.

On the other hand, quadrupole instruments, by the very nature of their mass filter action, are amenable to a wide range of SIM applications. Selected *m/e* values throughout the entire mass range can be put into focus readily and this type of instrument can bridge the entire gap from rapid cyclic scanning to SIM with no distinct dividing line between the two modes. To date, the manufacturers of these instruments have concentrated their programming efforts on these two extreme cases; however, nothing precludes their expanding the area of user-controlled operational modes anywhere in the continuum from the monitoring of one ion only to the monitoring of every ion within the complete range of the mass spectrometer. MIS-SIM programs are not currently commercially available for these instruments; however, such an application can be accommodated with ease on quadrupole mass spectrometers.

In the unlikely event that all of the instrumental and sampling problems associated with GC-MS can be controlled to the point that only the ion statistics are accuracy limiting factors, a single, clearly defined

Table IV

Summary comparison of MIS-SIM and MSSMET techniques

MIS-SIM	MSSMET
High precision	Medium precision
Linear range $>10^4$	Linear range $\approx 10^3$
Data collection requires operator intervention for optimum results	Highly automated data collection
Semi-automated quantitative analysis	Fully automated quantitative analysis
Manual peak identification	Automated peak identification
Real-time retention index determination	Automated off-line retention index determination
Up to about 10 compounds monitored	All compounds monitored
Isotope carriers needed for maximum sensitivity	Isotope dilution technique does not significantly extend range

method would still not evolve. The variation in analytical need will continue to encourage variation in the analytical method. The detector has only a limited time to observe each GC peak as it appears, rises and falls. Should it spend a large amount of time on a few peaks or a small amount of time on many peaks? Modern instrumentation will give the individual scientist the power to divide this time up as he sees fit. If the past is any indication of the future, he will see an unlimited number of ways to do so.

ACKNOWLEDGEMENTS

The authors wish to acknowledge support for this undertaking from an NIH Research Grant (RR-00480) and an NIH Postdoctoral Fellowship award to Dr. R.H. Thompson.

REFERENCES

1 R.S. Gohlke, *Anal. Chem.*, 31 (1959) 535.
2 R.A. Hites and K. Biemann, *Anal. Chem.*, 42 (1970) 855.
3 Hewlett-Packard HP 5933A Mass Spectrometer Data System.
4 S.C. Gates, N.D. Young, J.F. Holland and C.C. Sweeley, in A. Frigerio and N. Castagnoli (Editors), *Advances in Mass Spectrometry in Biochemistry and Medicine*, Vol. 1, Spectrum Publications, Holliswood, N.Y., 1976, Ch. 44, p. 483.

5 C.C. Sweeley, N.D. Young, J.F. Holland and S.C. Gates, *J. Chromatogr.*, 99 (1974) 507.
6 S.C. Gates, N.D. Young, J.F. Holland and C.C. Sweeley, in A. Frigerio and D.M. Desiderio (Editors), *Advances in Mass Spectrometry in Biochemistry and Medicine*, Vol. 2, Spectrum Publications, Holliswood, N.Y., in press.
7 S.C. Gates, M. Smisko, J.F. Holland and C.C. Sweeley, in preparation.
8 F. Kovats, *Helv. Chim. Acta*, 41 (1958) 1915.
9 M. Axelson, T. Cronholm, T. Curstedt, R. Reimendal and J. Sjövall, *Chromatographia*, 7 (1974) 502.
10 F.W. McLafferty, R.H. Hertel and R.D. Villwock, *Org. Mass Spectrom.*, 9 (1974) 690.
11 M.G. Horning, J. Nowlin, C.M. Butler, K. Lertratanangkoon, K. Sommer and R.M. Hill, *Clin. Chem.*, 21 (1975) 1282.
12 D. Henneberg, *Z. Anal. Chem.*, 183 (1961) 12.
13 C.C. Sweeley, W.H. Elliott, I. Fries and R. Ryhage, *Anal. Chem.*, 38 (1966) 1549.
14 J.F. Holland, C.C. Sweeley, R.E. Thrush, R.E. Teets and M.A. Bieber, *Anal. Chem.*, 45 (1973) 308.
15 N.D. Young, J.F. Holland, J.N. Gerber and C.C. Sweeley, *Anal. Chem.*, 47 (1975) 2373.
16 S.C. Gates, R.H. Thompson, J.F. Holland and C.C. Sweeley, in preparation.

EVALUATION BY MASS FRAGMENTOGRAPHY OF METABOLIC PATHWAYS OF ENDOGENOUS AND EXOGENOUS COMPOUNDS IN EUKARYOTE CELL CULTURES

PRUDENT PADIEU and BERNARD F. MAUME

ERA CNRS 267: Laboratoire de Biochimie Médicale (Faculté de Médecine) et Laboratoire de Biochimie des Interactions Cellulaires (Faculté des Sciences de la Vie et de l'Environnement) de l'Université de Dijon, Dijon (France)

SUMMARY

The most generalized method for the assay of tissue metabolites is classically based upon enzymic reactions which are monitored by liquid-phase analysis. In many methods, no separation is carried out. Such a method is ruled out if analytical parameters are not specific for a unique compound in the liquid phase. For instance, in the enzymology of hormone metabolism these enzymic assays are not amenable to the study of the metabolic fate of compounds often in great number and at minute levels. Often it is not known to what extent the enzymes are substrate specific. For instance, in the liver, the reductive catabolism of corticosterone leads to 14 compounds for which their respective productions are linked to sex and age, and a position isomer of corticosterone. 18-hydroxy-11-deoxy-corticosterone, follows the same reductive route. In adrenals some reduced metabolites arise from these two steroid hormones and are age dependent. In liver, the metabolites from a single xenobiotic compound are exceedingly numerous and various. It is necessary, then, to use separative methods to isolate the many metabolites produced. When such metabolites are amenable to volatilization for gas-phase analysis, the interfacing of the gas chromatograph to the mass spectrometer allows one to identify each compound introduced into the spectrometer. Among the ions produced by fragmentation of a compound or of a family of compounds, several specific fragments can be selected to be recorded along the chromatographic run leading to mass peaks which are quantitatively proportional to the amount of each compound, provided other foreign molecules do not contribute to the production of the same fragments. These methods, called mass fragmentography or multiple-ion detection, or selected-ion recording, allow, with all the resources of gas chromatography such as sample derivatization with stable-isotope labeled reagents and the use of the same unlabeled derivatized molecules as carriers and standards at once, quantitation at the level of a picomole or lower. Examples will be given of the study of the metabolism of hormonal steroids and of safrole, a carcinogenic compound, by differentiated eukaryotic cells in culture from the rat.

INTRODUCTION

Nowadays, under the impulse of molecular biology, it is undeniable that biochemistry, physiology, microbiology, virology, genetics, biophysics and physical chemistry have merged into a single science in the service of cell biology. The actual problem encompassed by these sciences is the understanding of factors that act on the organization of cells as living units, tissues and organisms with the concepts of the leading role and the totipotentiality of desoxyribonucleic acid (DNA) within each cell[1]. It is necessary to study in detail the mechanisms that regulate the action of genes in a cell during its development and in relation to its environment. Progress in this field depends on the choice of the biological system under experimentation and the ability to perform significant analyses in the whole animal as well as in systems *in vitro*.

Among biological systems *in vitro*, tissue cultures were thought at its very outset to replace true living systems such as animals, surviving organs, organ slices or organ homogenates. Cell cultures are essentially a new type of laboratory animal, and their uses in the field of developmental biology were marked by success but also by various pitfalls. Difficulties were encountered, first in obtaining well-defined and stabilized cell types in culture, and second in performing quantitative assays on such small tissue samples since a regular 21-cm^2 culture dish usually contains only few milligrams of fresh cells.

The cellular activities of the culture are regulated by trophic factors, *i.e.* nutrients and hormones as far as specific enzymic and hormonal targets are present in the cells; therefore the following two kinds of factors may act within such cultures.

(1) Extracellular factors: energetic nutrients, plastic nutrients for macromolecular syntheses, among which vitamins, hormones and some

nutrients are essential. Most of these nutrients are supplied by the synthetic medium except for lipids, hormones and more or less known factors supplemented by the sera added to the synthetic medium.

(2) Intracellular factors restricted to gene action because their relations to extracellular factors are frequently ill defined.

Therefore, studies that can be carried out with the help of cell culture include the three principal aspects of cell biology.

(1) Cell differentiation which covers the embryological development of gene expressions and then the control of their expressions in the mature cell including the role of extracellular factors: nutrients, hormones, xenobiotic compounds and virus.

(2) Cell genetics, which are essentially molecular genetics of normal cells and mutant cells owing to physical, chemical and viral factors to study the mechanism of induction and repression in relation to the reciprocal influences between the cellular genetic information and the mutation factors.

(3) Cell physiopathology of either inherited molecular diseases in cultured cells from pathological tissues or of metabolic disorders induced in normal cultured cells.

Since the initial discovery by Rous and Jones[2] in 1916, revived in 1952 by Moscona[3], trypsin or other proteolytic enzymes are used to dissociate a tissue into isolated cells to allow inoculation in a suitable culture medium which can sustain cellular growth and proliferation. Important progress has been made in the culture of normal isolated cells from fetal[4] or postnatal heart[5], liver[6-9], adrenals[10-12] and other tissues (as reviewed in refs. 13 and 14). These cultured cells offer an excellent model system for the investigation of many biochemical mechanisms among the different cellular functions, as mentioned above.

In the laboratory, eukaryotic cell cultures used are the following.
(1) Beating heart cells to study the regulation of macromolecular biosynthesis[15] and specific protein syntheses, for example of cardiac myosin and actin, and the occurrence of rare amino acids such as 3-methylhistidine in actin and N^{ε}-monomethyllysine in myosin[16].
(2) Liver-cell culture, to study the ontological and sexual differentiation of the metabolism of steroid hormones[9,17,18,30] and to elucidate the pathways and the enzymic inductions involved in the metabolism of a xenobiotic compound such as safrole: a model for compartmental liver-cell metabolism and carcinogenesis.

(3) Adrenal cells to study ACTH action on hormonogenesis and the role and fate of membrane receptors[11,12].

CONDITIONS INVOLVING THE JOINT USE OF CELL CULTURES AND GAS-PHASE ANALYSIS IN DEVELOPMENTAL AND CELLULAR BIOLOGY

Criteria for the use of cultured cells in developmental and cellular biology

(1) The obtaining of a homogeneous cell population either through the selective cellular growth from a mixed cell population or preferably by cloning cells from an initial mixed cell population.

(2) The maintenance of enzymic equipments corresponding to the retention of certain specific functions of the tissue explanted from the organism.

(3) The maintenance of such functions through a sufficient number of subcultures for time course studies which in addition must sometimes be associated as a prerequisite condition to the non-appearance of signs of cell transformation (ranging from the loss of euploidy to oncogenic transformation).

(4) The maintenance of regulating mechanisms through nutrient and hormonal actions including the conservation of membrane sites mediating induction-repression mechanisms.

(5) The existence of specific chemical and biochemical markers.

Necessities, conditions and advantages of gas-phase analysis[20]: gas-liquid chromatography (GLC) and mass spectrometry (MS) for metabolic studies with cell cultures

"It is now clear that gas-phase analytical work is emerging as a new field of chemistry ... well suited for the study of complex mixtures of organic substances on a microgram or submicrogram scale and they are of peculiar importance for work in the fields of biology and medicine ... The key to new discoveries lies in new methodology. It is to be hoped that the field of endocrinology will benefit from these advances which are based on the physical sciences but which find their most important use in biological problems". This quotation from E.C. Horning written in 1968[20] has kept its prophetic power about better achievement in cell-biology research and has gained more and more signification through the wide expansion of quantitative gas-phase analysis.

Obviously the development of gas-phase analysis in cell biology implies that biochemical compounds, after chemical derivatization, eventually must be volatile in order to be analyzed in the gas chromatograph and with the interfaced mass spectrometer (GC-MS). When this prerequisite is achieved, gas-phase analysis through GC-MS exhibits many advantages that make it ideally fitted to cell culture.

Enzymic assessment: liquid-phase analysis versus gas-phase analysis

Two procedures may be applied to the study of a multi-enzymic route: (1) the measurement of the activity of each enzyme in the cellular extract by the assay of coenzyme of compound productions, and (2) the quantitative analysis of all the substrates produced and utilized in the enzymic sequences.

Classical methods of enzymic assay. Obviously, in tissue culture, the first method is often prohibited not only owing to lack of the necessary cellular sample size but also owing to time consumption when several enzymes have to be analzyed. Another serious drawback for such enzymic analysis is that several enzymes are not absolutely substrate specific, for example some steroid reductases, and mono-oxygenases of the liver. In addition, the lack of sensitivity and of specificity of liquid-phase spectrophotometric analysis, and the difficulties of quantitative analysis in liquid chromatography, bring overwhelming difficulties that prohibit the use of these methods of classical enzymology. In liquid phase analysis, the obligatory solute-solvent interactions do not allow a mass assessment free of carrier interactions and they oblige one to measure physical or chemical parameters of the molecules. In these conditions, sensitivity and specificity shortcomings usually prevail except in radioimmunoassay if the antigen-antibody system is specific.

Analysis of substrates and products by gas-phase analysis[20]. Gas-phase analysis introduces separative methods in which there are no solute-solvent interactions. Therefore the solute can be measured in terms of its true present mass either by the common hydrogen flame ionization detection in GC or by ion detection in MS. These detectors are not specific and exhibit a wide linear response to the mass of each compound or fragment. The specificity of detection results only from the resolving power of the chromatographic and/or of the mass fragmentographic systems. Indeed this method of quantitative assay of each substrate or metabolite intervening in the multi-enzymic chains remains feasible even if two metabolites differ in a single stereomeric feature such as a hydrogen position. In mass fragmentography, the mass spectrometer becomes a specific mass detector for

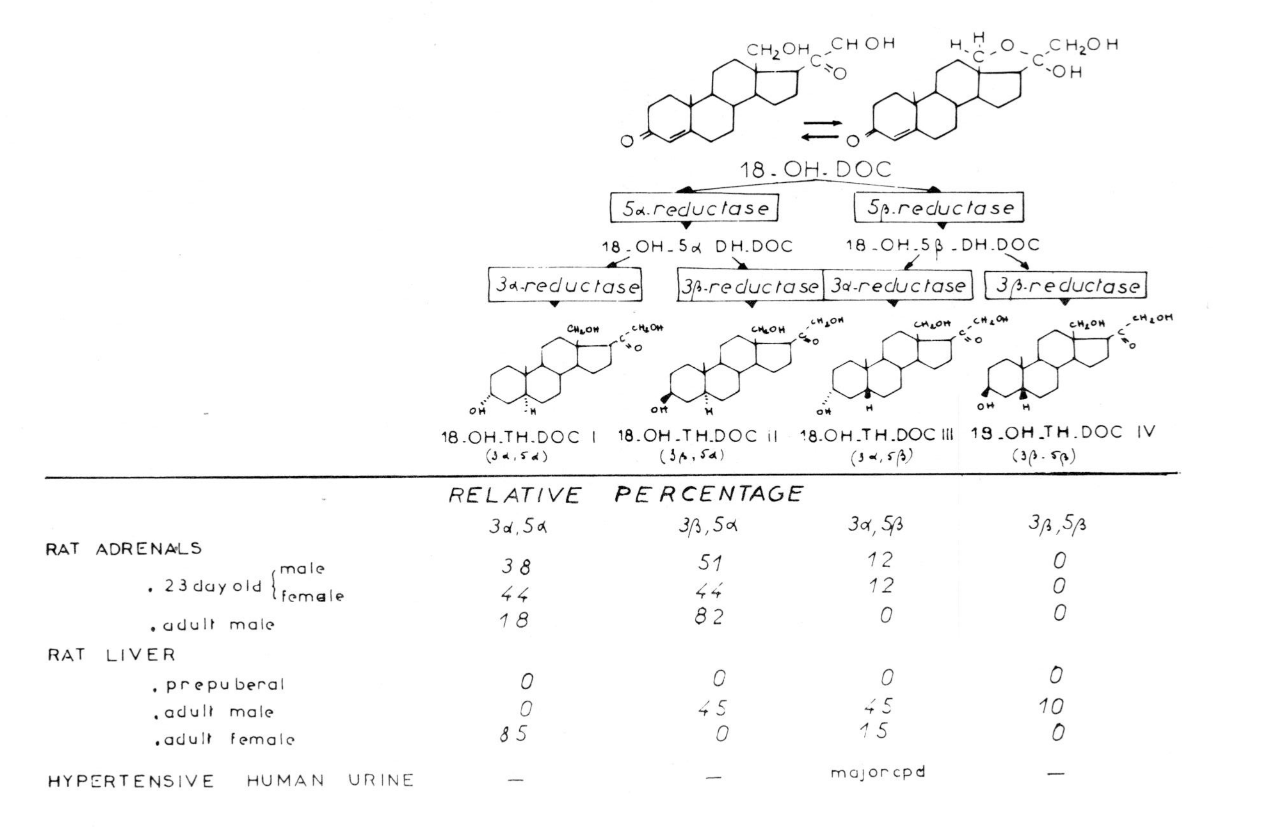

	RELATIVE PERCENTAGE			
	3α,5α	3β,5α	3α,5β	3β,5β
RAT ADRENALS				
. 23 day old, male	38	51	12	0
. 23 day old, female	44	44	12	0
. adult male	18	82	0	0
RAT LIVER				
. prepuberal	0	0	0	0
. adult male	0	45	45	10
. adult female	85	0	15	0
HYPERTENSIVE HUMAN URINE	—	—	major cpd	—

Fig. 1. Isomeric metabolites produced by the enzymic reductions of 18-OH-11-deoxycorticosterone, and their relative percentages in rat liver and adrenals (see ref. 55, 57). An analogous scheme is found for corticosterone in the same rat organs (see refs. 12, 30, 56).

quantitative measurement after it has been used as a means to identity by their mass spectra the structures of the different compounds to be quantitated.

With the mass spectrometer, generally in the GC-MF configuration, the biological laboratory rediscovers the mass measurement not in the earth's gravitational field with a balance but in other displacement fields, the magnetic field and/or electric field. This is a comeback to the golden rule of the analytical chemist: isolation, identification and mass measurement.

The few examples given below will show how GC-MF represents up to now the best analytical system. For instance, corticosterone and 18-hydroxy-11-deoxycorticosterone (18-OH-DOC) are position isomers and they can each give rise to fourteen metabolites by sequential reductions (Fig. 1):

(1) the double bond in position 4-5 gives two isomers, 5α and 5β;

(2) the oxo group in position 3 gives four isomers, 3α-β/5α-β, and the one in position 20 gives eight isomers, 3α-β/5α-β/20α-β.

Another striking example is given by the metabolism of safrole in the whole rat where there are over 50 compounds not yet completely identified, although 15 compounds are identified among the 20 found in liver-cell culture (see Figs. 11 and 12).

Tissue micro-sample analysis

The minute amounts of available tissue as already mentioned do not preclude the simultaneous analytical fractionation by GC of many metabolites of a compound family, their individual characterization by mass spectrometry and their quantitation by specific ion detection[21] in GC-MF.

Methodology refinement in gas-phase analysis

Most of the analytical problems in the fractionation of volatile compounds can be solved owing to the many species of stationary phase, the many types of chemical derivatization, the use of packed and capillary columns and preliminary fractionation by liquid chromatography[22,32]. In mass spectrometry methodology the combined use of electron-impact ionization[23], chemical ionization[24], field ionization[25] and field desorption[26] at low pressure or of atmospheric pressure ionization[27] as well as the isotopic labeling for analysis without biotransformation[28] or for tracing metabolic pathways[29,30], allow one to increase both the specificity and the sensitivity of mass spectrometric identifications and of mass fragmentographic assays. In addition, the different types of mass spectrometer (the low-resolution mass spectrometer either with a magnetic field sector or with monopolar or quadrupolar radio frequency and electric fields or the high-resolution mass spectrometer), in association with the different modes of gas chromatography, ionization and fragmentation, offer a wide possibility of combination to solve any problem while the main drawback may be financial stringency.

Methodological improvements and potentialities arising from the association of tissue culture with gas-phase analysis

The advantages of the joint use of cell culture and gas-phase analysis for quantitative analysis are the following.

Constancy of the extra-cellular space and metabolism by a single tissue. Cells in culture are confined in a constant extracellular space which can be sampled and modified at will. Extracellular space and intracellular space constitue a biphasic system from which any other compartments can be defined when their analytical discriminations are feasible. This is the most important advantage of tissue culture *versus* whole animal where any injected substrate and resulting metabolites undergo a continuous dilution and excretion in many compartments. Consequently, the study of the metabolism of a compound in cultured cells can be referred to a unique tissue in a steady biphasic system, whereas in the organism the same compound, *e.g.* a steroid, an amino acid or a lipid, may be metabolized in several places into different compounds which may be re-utilized.

Most of the metabolisms are associated with metabolite transports through cell membranes. These transports are more easily investigated in the culture dish. In addition to the study of transport mechanisms, cellular metabolisms can be studied only by extracting the compounds from the culture medium space without destroying the culture. Therefore, the culture can be re-utilized for other experiments after being rinsed and re-adapted to the standard culture medium.

Tissue sample reproducibility. From a single inoculum of cells many reproducible units, *i.e.* many culture dishes, can be prepared for time course studies, as well to multiplicate each experimental point. Since cells can be kept frozen in liquid nitrogen for several years and then cultured again after thawing, experimentation can be repeated at any time or pursued with the same biological material but under other parametric conditions.

Metabolic studies with isotope labeled substrates. The absence of extracellular space dilution renders the use of labeled compounds efficient, especially when isotopic ratio measurements are performed by mass fragmentography. Stable isotopes (D, ^{13}C, ^{15}N, ^{18}O, etc.) as well as ^{14}C--labeled compounds can be used. Radioactive ^{14}C-compounds can be added up to 1 μCi to 5 ml of culture medium without endangering the cells by radiolysis, but usually 0.1 μCi per 5 ml of medium in a 21-cm^2 dish is ideally suitable for GC-MF analysis.

^{14}C-compounds are more useful in tissue culture studies for the following reasons[30,31].

(1) Many ^{14}C-compounds are available compared with the scarcity and the cost of ^{13}C-compounds.

(2) Preliminary TLC separation of metabolites and radioscanning of the plate allow one to locate the metabolites and to make a pre-purification before GC-MS and GC-MF analyses[32].

(3) Most of the ^{14}C-compounds can be incubated with the cultured cells in physiological amounts without dilution of the commercial radioactive sample[30]. For instance, for several steroid hormones, the normal physiological levels in the plasma range from 0.02-0.2 nmole/ml. Consequently, for 5 ml of medium per dish the incubation of 0.1-1 nmole of compound with a specific activity of 50 mCi/mmole given by one ^{14}C labeling will result in the addition of 10^4-10^5 dpm in each dish.

Table I

Natural heavy isotope contribution to the M+2 molecular ion by measurement of the (M+2)/M intensity ratios on a 10-ng testosterone derivative (MO-TMS) by multiple-ion detection (MID) and multiple-ion detection-peak matching (MID-PM)

	$\frac{M+2}{M}$ x 100	Standard deviation
Calculated	8.52	
MID-PM	$9.34 \pm 0.08^{+}$	0.9%
MID alone	$10.5 \pm 0.5^{++}$	4.7%

$^{+}$Mean of 12.
$^{++}$Mean of 5.

Table II

Comparison between isotopic ratios calculated from the known specific activity (5o mCi/mmole) and measured by MID-PM

For both methods, MID and MID-PM, the ratio 391/389 is corrected for the contribution of natural isotopes (^{30}Si, ^{13}C, ^{15}N, ^{18}O) for 391 value and for the background effect.

$^{14}C/^{12}C$ calculated from the SA of 4-^{14}C-testosterone	391/389 measured by MID-PM
0.24	0.24
0.26	0.30
0.32	0.36
0.37	0.37
0.67	0.68
1.25	1.31

In addition, such isotope ratios are easily mesurable by GC-MF. Since the absolute specific activity of one ^{14}C is 63.5 mCi/mA and account being taken of the contribution of the M+2 natural isotopes, Table I gives an example of the ^{12}C-testosterone-3-MO-17-TMS where for the M+2 ion the calculated natural percentage of M+2 isotope, 100 (M+2)/M = 8.52. is compared with the values measured by either the multiple-ion detection-peak matching (MID-PM) method[33] (9.34 ± 0.08) or by simple multiple-ion detection (MID)[28] (10.5 ± 0.5). Table II shows the comparison of isotopic ratios $^{14}C/^{12}C$ for 4-^{14}C-testosterone-3-MO-17-TMS measured by MID-PM and compared with the calculated value from the specific activity of 50 mCi/mmole and after dilution. Such a method of $^{14}C/^{12}C$ isotopic ratio measurement can be realized down to 3-5%. For example, the assay of 10 ng (30 nmoles) of 4-^{14}C-testosterone with an isotopic ratio of $^{14}C/^{12}C$ = 10 corresponds approximately to the assay of 40 dpm by β-scintillation counting. Nevertheless, the mass spectrometer cannot replace the scintillation counter when there exists a high isotopic dilution even with a high absolute radioactivity.

Another important advantage of ^{14}C-compounds over ^{13}C-compounds in tissue culture is that the labeling by one atom per molecule will produce a M+2 isotopic ion or fragment which will be easily detected and measured since the natural M+2 isotopic peak is less important than the M+1 natural peak. Therefore D- or ^{13}C-labeled compounds must contain at least two isotopic atoms. This is a factor of increasing difficulty and cost of synthesis, especially for ^{13}C, and in addition deuterium should be incorporated into the molecule where there is no risk of protium exchange.

Use of cultured cells to synthesize intermediate metabolites. The elucidation of a metabolic sequence is greatly facilitated when intermediate compounds are incubated instead of the precursor. In animal studies such elucidation is a difficult undertaking since the compounds are seldom commercially available, are expensive and are difficult to synthesize in a labeled form. They can be biosynthesized by a batch of cultured cells incubated with the available labeled precursor and then fractionated and purified by suitable chromatographic methods. Since most metabolic studies by MF can only be performed on a few 21-cm^2 dishes (3-10) such production of a wide variety of labeled intermediates is only feasible in amounts which of course would be insufficient for animal or conventional experiments *in vitro*.

QUANTITATION BY MASS FRAGMENTOGRAPHY ANALYSIS (GC-MF)

Development of specific ion detection (Fig. 2)

Mass fragmentography, a term coined by Hammar and Hessling in 1968[34] in the course of studies on drug metabolism, is also called multiple-ion detection (MID) or selected-ion recording or monitoring. The recording of a single ion (or single-ion detection, SID) was proposed for the first time by Henneberg in 1961[21], and the simultaneous recording of several ions (MID) was proposed by Tal'Rose *et al.* in 1965[35]. In 1966, Sweeley *et al.*[28] gave the first example of the use of an acceleration voltage alternator (AVA) for the resolution of a mixture of epiandrosterone and dehydroepiandrosterone by MID and for the discrimination between a natural compound and its deuterium-labeled species (the penta-O-trimethylsilyl derivatives of glucose and d_7-glucose). Brooks *et al.* in 1971[36] described the use of SID for the analysis of steroidal drugs, and Baillie *et al.* (1972) for 11-deoxycortisol as a model compound[37]. In the field of catecholamines, Koslow *et al.*[38] described in 1972 an MF assay of norepinephrine and dopamine in the brain. In drug metabolism this continuous monitoring of several ions has been used by Gaffney *et al.*[39].

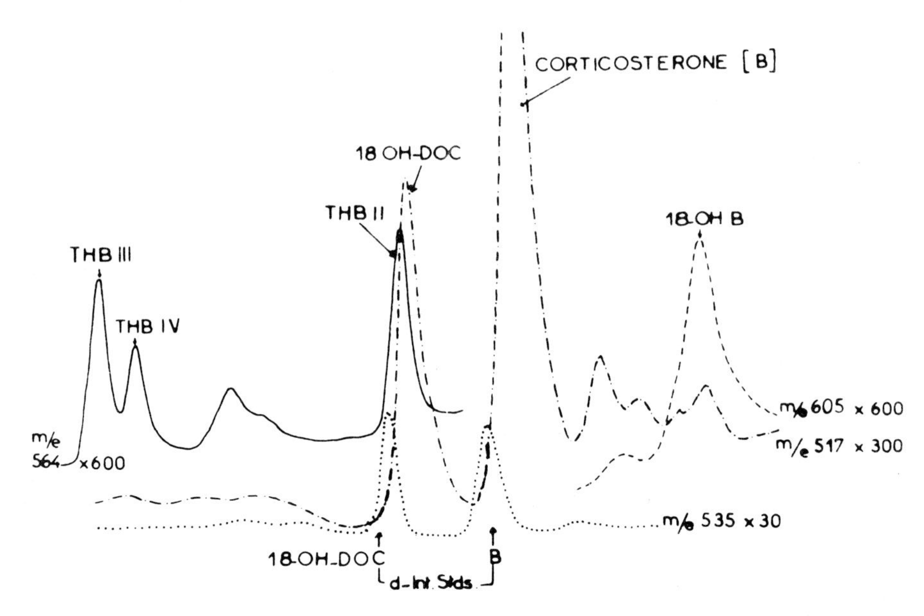

Fig. 2. Mass fragmentogram from 4-day-old male rat adrenal extract in which MO-TMS-steroids were assayed by specific ions: *m/e* = 517 for corticosterone (B) and 18-hydroxy-11-deoxycorticosterone (18-OH-DOC), *m/e* = 564 for tetrahydrocorticosterone (THB) isomers) (see Fig. 1 for structures), *m/e* = 605 for 18-hydroxycorticosterone (18-OH-B) and *m/e* = 535 for d_{18}-18-OH-DOC as perdeuterated carriers and internal standards (see ref. 54).

The interfacing of GC-MS with a computer

In 1970, Hites and Biemann[40] proposed the repetitive scanning by a computer of the mass spectra from the GC effluents, and Sweeley *et al.*[41] described the computerized data acquisition. The repetitive scanning during GC-MS runs with an off-line computer was reported by Reimendal and Sjövall in 1972[42]. The use of a computer to collect, store and process the data extends the scope of the analyses, increases the output of the system since it abolishes the bottleneck resulting from the reduction of the spectra, and enhances the accuracy and the reliability of the methods. Moreover, the storage of data not immediately exploited is a very important aid in the retrospective interpretation and elucidation of results such as the description of new metabolites. Such computerized methods were developed for MF analyses in 1973 by Holland *et al.*[43], Baczynskyj *et al.*[44], Holmes *et al.*[45] and Watson *et al.*[46] for magnetic field mass spectrometers and in 1974 by Caprioli *et al.*[47] for a quadrupole mass spectrometer. With this type of mass spectrometer, the voltage changes can focus selected ions throughout the entire mass range of the mass spectra. But as in other MS, the number of monitored ions during a cycle remains limited by the peak elution time during which monitoring cycles should be repeated enough to collect accurate statistical data. Recently, Young *et al.*[48] have described the control by the computer not only of the accelerating voltage but also of the magnetic field for multicomponent analyses with a magnetic mass spectrometer.

Use of isotopic labeling for quantitative mass fragmentography analyses

Since the first use of a deuterium-labeled compound in MF analysis by Sweeley *et al.* in 1966[28] a new approach involving the dilution of the biological sample with the stable isotopic labeled compound that serves both as carrier and standard has enabled isolation and quantitation of compounds such as prostaglandins by Samuelson *et al.*[49] and Axen *et al.*[50], insect juvenile hormone by Bieber *et al.*[51], 5-hydroxyindole acetate by Bertilsson *et al.*[52], drugs by Gaffney *et al.*[39], the biosynthesis of steroid hormones in rat adrenal by Maume and coworkers[53-55], the metabolism of corticosterone and of 18-OH-11-deoxycorticosterone in rat liver by Bournot and coworkers[56,57], the metabolism of steroids in rat-liver cultured cells such as testosterone and corticosterone by Bournot *et al.*[30] and Chessebeuf *et al.*[9,58], and the biosynthesis of corticosteroids in adrenal cell culture by Maume and coworkers [11,12,59].

Mass fragmentographic utilizations and methods are at present in an exponential development, and it is outside the scope of this article to review all their improvements and applications. Readers may find many quotations of papers dealing with mass fragmentography as well as with GC-MS in the following compilations and the symposium books[60-66].

Use of standards

Quantitation with high sensitivity and specificity is the most interesting feature of mass fragmentography. To reach the nanomole, the picomole or even a lower level, the use of standards is an absolute requirement, especially when a standard can also act as a carrier to prevent compound losses in the whole GC-MS tubing.

Types of standard

The careful choice of the primary standard which is added at the earliest possible stage of the extraction of sample or tissue and of the secondary standard which is added when the sample is treated for gas-phase analysis are important steps for both quantitations in the sample in respect to compound yields and in the final volume injected in the GC-MS in respect to peak evaluation. The primary standard can be a parent radioactive compound, not present in the sample, or traces of ^{3}H-labeled compound if ^{14}C is the metabolic precursor. The remaining radioactivity will allow calculation of loss during all the steps leading to the final stage when the sample is ready to be derivatized before or after addition of the secondary standard.

Secondary standards may take one of the following forms

1. A non-biological compound of similar chemical nature with either a different retention time or a different monitored ion.

2. A non-biological structural isomer of the sample which must have a different retention time if it will produce iso-fragments.

3. An isotopic form of the sample depending on which of the sample or the standard is the labeled molecule and of the type of label.

Forms 1 and 2 are the least satisfactory since they cannot be used as carriers if they have, or they must be chosen with, a different retention time. However, they appear useful and economical when the help of a carrier is not needed. Examples are given by the use of tetrahydrocortisone in studies of the metabolism of corticosterone in rat since this animal does not produce 17α-OH-corticosteroid owing to the lack of 17α-steroid hydroxylase[56], the use of corticosterone in the study of tetrahydro derivatives of 18-OH-11-deoxycorticosterone in rat liver[57], of another steroid in the assessment of estriol in pregnancy urine[67], or 16-epi--estriol for the assay of estriol and 17β-estradiol in non-pregnant women[53] and drugs in the determination of barbiturates in plasma samples[68]. Form 3 appears to be the best standard and is an almost ideal one since its retention time is almost the same.

Stable isotope-labeled molecules

Albeit the heavy-isotope-labeling peak is seen on the mass chromatogram with a retention time shorter by a few msec, a single peak still appears on the gas chromatogram or on the total ion current recording. This cannot be true if the mass increase is relatively important in relation to the mass of the unlabeled compound and especially if a capillary column is used. The utilization of such stable isotope compounds is the basis for a stable-isotope dilution method which can be done in different ways.

Metabolic precursors. The stable-isotope-labeled compound can be the metabolic precursor. The isotopic ratio assessment of each metabolite, as far as the label is retained or lost by the molecules allows one to establish the metabolic routes and to calculate turn-over rates as well as net syntheses. Unless the legal regulations for drug administration to human beings forbids, stable-isotope-labeled molecules may be used in human investigations with precaution and under ethical rules. In our country the French Health Laws positively forbid, among many radioactive isotopes, the use of ^{3}H and ^{14}C in human experimentation. With stable isotopes, many fruitful metabolic studies have already been carried out in man[29,62].

Stable-isotope-labeled carrier standards. One of the first examples was given by Samuelson *et al.* in 1970 [49] for the assay of prostaglandin PGE_1 at the nanogram level using the perdeuterated methyl group in a methoxime derivative (d_3-MO). This is the type of secondary standard use that does not preclude the loss of PGE_1 during the tissue sample extraction if a primary standard is not associated. Then and now several examples are given by Maume *et al.*[53,69], by Bournot *et al.*[30,56], and Prost *et al.*[54,55] of the use of perdeutero-trimethylsilylation (d_9-TMS) of steroids for quantitative analyses of corticosteroid metabolism in rat liver and adrenals during ontological development. The derivatization in d_9-TMS-ethers was used for the first time by McCloskey *et al.*[70] as a tool for elucidation of structure by mass shifting and comparison of fragment values between the normal and the labeled compound.

The reverse way: labeled sample and unlabeled carrier derivatizations. In spite of the assertion that intramolecular labeled carrier compounds are preferable, the labeling of compounds through derivatization is the most useful way when such labeled compounds are not available. There are obvious reasons to keep the isotopic derivatization for the sample to be analyzed and to use the same protium-derivatized molecule as the secondary standard and the carrier in the necessary amount (100-500 times the sample) to prevent losses during the analysis. The advantages are the reduction of the cost of the experiment and the increase of the accuracy of the mass fragmentography as set forth in the laboratory by Bournot[71] and by Maume *et al.*[12]. The deuterated derivatization (mostly d_9-TMS) of the sample lessens the inaccuracy of the isotopic ratio measurement. Indeed the labeled reagent is never 100% pure. But if we suppose its purity is 99%, and the biological compound is 1 ng whereas 100 ng will be used as a carrier and secondary standard, the formation of 1% of unlabeled compound in the deuterated biological sample (*i.e.* 0.01 ng) will be absolutely negligible in regard to the 100 ng of carrier which is the same compound but protium derivatized. In the classical method, the addition of 100 ng of deuterium-derivatized carrier will carry 1 ng of protium-derivatized sample and consequently an error of 100%. If the protium-derivatized carrier is quantitatively added to the deuterium-derivatized sample the two peaks can be measured accurately by comparing the records at x1 and x100 sensitivity. In addition, a less expensive radioactive labeling of the same or of a parent molecule will be used as the primary standard such as the ^{3}H- or the ^{14}C-compound in trace amounts. The loss of sample during extraction is easily assessed by scintillation counting. In addition, when TLC prepurifications are carried out, the radioscanning of the TLC plates will greatly help the subsequent gas-phase analysis. The radioactive primary standard, if it is the same molecule, must not bring a significant dilution of the sample.

Double labeling for metabolites studies by gas-phase analysis

When experiments are carried out with animals or cell culture the use of ^{14}C-substrate will allow one to carry out isotopic ratio measurements of the ^{14}C-substrate and its metabolites by MF. This method is ideally suited for cell culture and experimentation *in vitro*, since the initial specific activity of the substrate can be adjusted to obtain metabolites with suitable isotopic abundance. As already shown in Table I the M+2 abundance of the natural and the labeled molecules can be calculated by MID-PM with a standard deviation of less than 1%. Therefore the measurement of the $^{14}C/^{12}C$ isotopic ratio can be carried out down to 3-5%, the amount of corresponding radioactivity assessed being only dependent on the sensitivity of the MF method which is at present in the same range as the β-scintillation counting. Each metabolite being separated, the ratio $^{14}C/^{12}C$ of the two mass peak areas gives the specific activity, and the sum of the two areas measures the net synthesis or transformation or transport of every compound, *i.e.* their compartmental distributions.

In a metabolic sequence, the comparison of the different metabolites through their structure and their specific activity value allows one to place them in a suite of reactions and to locate branched routes. Contributions by substrate already present in the biological system can be suspected when discrepancies in the specific activities (*i.e.* vicarious production or utilization of intermediary metabolites) do not fit with the postulated sequence of biochemical events. Conversely a homogeneous production of metabolites in respect to the specific activities found will help in establishing the sequence of biochemical reactions.

Mass fragmentography methodology reaches the goal of quantitative cell biology through an outstanding versatility emphasizing the predictive writing of E.C. Horning[20].

EXAMPLES OF METABOLIC STUDIES USING CELL CULTURES AND GAS-PHASE ANALYSIS

Already published work will be given as examples of the application of MF to cell biology studies in the field of the expression of cellular differentiated functions in tissue culture.

Steroid hormone biosynthesis and metabolism in normal rat adrenal cell cultures (Maume et al.[11,12])

Cells isolated from adrenals of normal post-natal rats are grown in monolayer cultures according to a method adapted from Chessebeuf *et al.*[9] using fractionated trypsinization for cell explantation to promote the selective growth of adrenal cells (spongiocytes) over fibroblastoid cells. Such primary cultures have been replicated with success. The metabolic pathways described herein have been assessed in primary culture and in two successive replicating subcultures using the formation of the d_9-TMS-ether for the biological sample and of the same unlabeled TMS compound as the carrier and the secondary standard as shown in Figs. 3-6. The principal metabolic features found in these cells are:

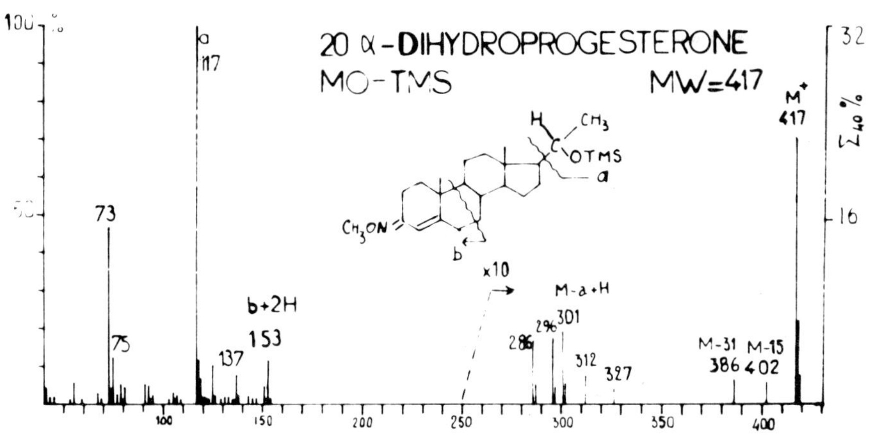

Fig. 3. Mass spectrum of 20α-dihydroprogesterone-MO-TMS, the reference compound used as a carrier and secondary standard for the identification of the metabolite (Fig. 5) found in rat adrenal cell culture incubated with ACTH and 4-^{14}C-progesterone by comparison with the mass spectrum of the MO-d_9-TMS derivative in Fig. 4.

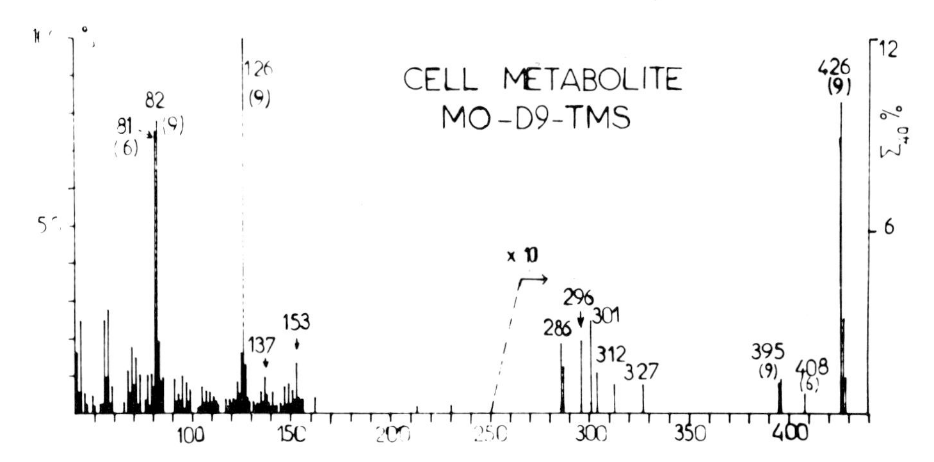

Fig. 4. MS of the MO-d_9-TMS metabolite found in adrenal culture and identified as 20α-dihydroprogesterone by comparison with MS in Fig. 3 and using d_9-TMS derivatization of the culture extract.

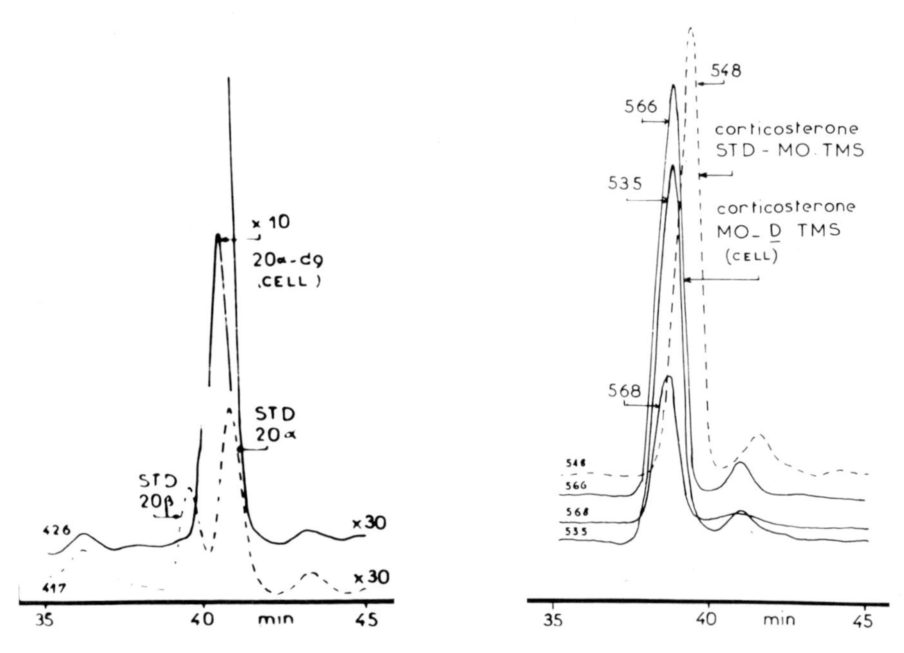

Fig. 5 (left). MF identification and assay of MO-d_9-TMS steroids of the adrenal culture extract using the same non-deuterated derivative as carrier and standard. Steroids: 20α-d_9 = 20α-dihydroprogesterone-MO-d_9-TMS; STD 20α = 20α-dihydroprogesterone-MO-TMS as internal standard.

Fig. 6 (right). Corticosterone under the same conditions.

(i) corticosteroids are produced only under ACTH stimulation;

(ii) the major normal corticosteroids *in vivo* are also produced in culture-corticosterone (7 μg/24 h/21-cm^2 dish), 18-hydroxy-11-deoxy-corticosterone (18-OH-DOC) and aldosterone;

(iii) adrenal fibroblastoid cells do not produce corticoids even under ACTH stimulation.

The use of thin-layer chromatography (TLC) and the radioscanning of the plates has demonstrated that 4-^{14}C-progesterone (533 ng (SA 50 mCi/mmole)/ 21-cm^2 dish) incubated with the culture medium alone was not metabolized but that progesterone was metabolized into 20α-dihydroprogesterone by adrenal fibroblastoid cells and that ACTH had no stimulating effect. In true adrenal cells in culture the metabolic transformation of progesterone into 20α-dihydroprogesterone occurred, but under stimulation by ACTH the bio-synthesis of corticosteroids was carried out as *in vivo* from acetate or cholesterol without the need of a steroid precursor adjunction in the culture medium. This is shown in Figs. 5 and 6 which also illustrate the method of deuterium derivatization of the biological sample of 20α-dihydro-progesterone and of corticosterone and the use of the unlabeled derivatives of 20α-dihydroprogesterone and of corticosterone which were added as secondary standards.

These investigations clearly show that isolated adrenal cells are able to grow in a cellular monolayer which retains the metabolic pathways for corticosteroid biosynthesis under stimulation by ACTH added to the culture medium. In association with MF methodology these cultures represent an efficient tool for studying cell growth and differentiation through hormonal regulation involving hormonal mediators. It is interesting to compare these cultures with a cell line from adrenal tumor[72] which does not synthesize corticosterone owing to the lack of 21-hydroxylation[73].

Metabolism of testosterone by liver cell lines

Physiological amounts (0.1 nmole per ml of medium) of 4-^{14}C-testo-sterone (SA 50 mCi/mmole) were incubated for 72 h with different cell lines [9,30]. After extraction, a pre-purification was done by TLC. Then TLC radioscanning allowed us to locate the radioactive areas, which were scraped off and eluted. MF analyses of MO-TMS derivatives were carried out on the ^{12}C-, ^{14}C- and d_9-TMS compounds. Each compound was identified by its GC retention time, its MF responses to at least two selected masses and the ratio of these two MF mass peaks in comparison with the standard. A change of the selected mass peak ratio will signify that impurities had been monitored at these masses in the cell culture samples.

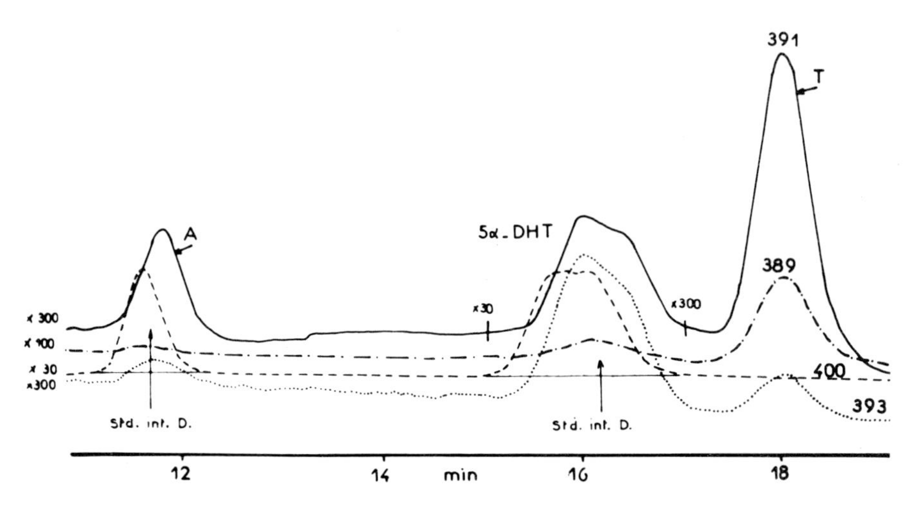

Fig. 7. Fragmentograms of compounds in two TLC overlapping spots. See text for explanations.

Fig. 7 shows fragmentograms of compounds in two TLC overlapping spots demonstrating the metabolism of testosterone (T) by cultured liver cells into two compounds: 5α-dihydrotestosterone (5α-DHT) and androsterone (A). The following masses of the MO-TMS derivatives of the molecular ions were observed: 389 for T, 391 for 4-^{14}C-T, 5α-DHT and A, 393 for 4-^{14}C-5α-DHT and 4-^{14}C-A and 400 for perdeuterated internal standards. In other cases M-31 fragments were the second selected masses for MF. Fig. 8 describes the metabolic pathways found in liver cell lines. Metabolic studies with sexed cell lines showed that adult male or female liver cell lines exhibited sex-linked testosterone metabolism which is the expression of phenotypic characteristics retained in these established cell lines[74]. Quantitation of metabolites under normal culture conditions and under the action of effectors has been carried out[74].

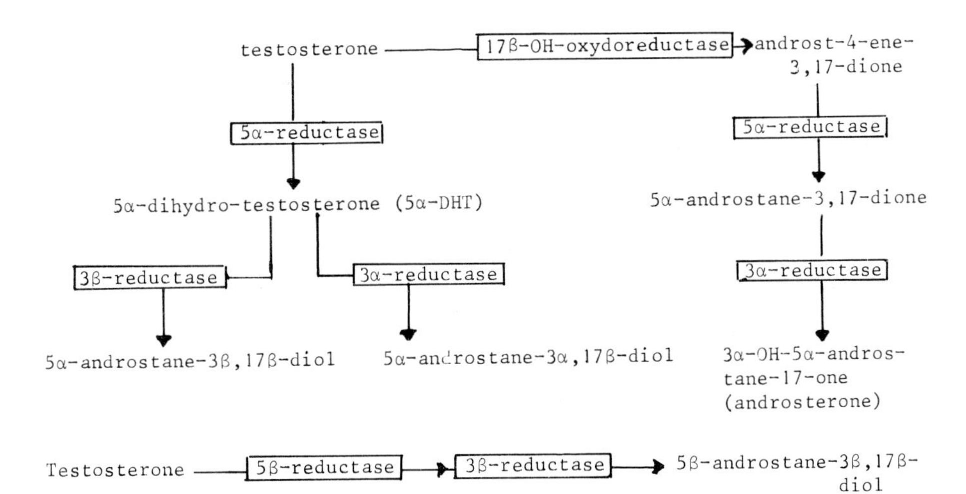

Fig. 8. Metabolic pathways of testosterone incubated with two post-natal rat-liver cell lines: 8FR and 18FR at different replications: 7th, 11th, 17th.

Metabolism of progesterone by fetal rat-liver and post-natal rat-liver cell lines

Liver is known to be the main site for progesterone metabolism [22,75,77]. During pregnancy, both fetal and maternal livers produce metabolites. Therefore, fetal, sexed post-natal and sexed adult liver cells in culture appear ideally suitable for the differentiation of the fetal contribution within the mother's urinary metabolites of progesterone. Work by Desgrès *et al.*[17,18] produced evidence for, among several well known metabolites of progesterone, the production of a 6α-OH-metabolite: 3β,6α-dihydroxy-5α-pregnan-20-one by fetal liver cells in culture which was the main metabolite (*ca.* 60% of incubated progesterone).

Fig. 9 shows the metabolic pathway found in these liver cell lines. Fig. 10 describes how, through the help of the 4-^{14}C-progesterone incubated within the cells, the labelling which was found of course in 3β,6α-dihydroxy-5α-pregnan-20-one, allowed one to identify this metabolite by GC-MS since doublet peaks from ^{14}C and ^{12}C fragments depended on the fragmentation. The 6 position of the hydroxy group was therefore assigned without ambiguity, and the 6α-hydroxy isomer was deduced by comparison with 3α,6α-dihydroxy-5β-pregnan-20-one and with mass spectra already published[77,78]. The hydroxylation in the 6α position was the main metabolic reaction in these two liver cell lines originating from 20-day-old fetuses

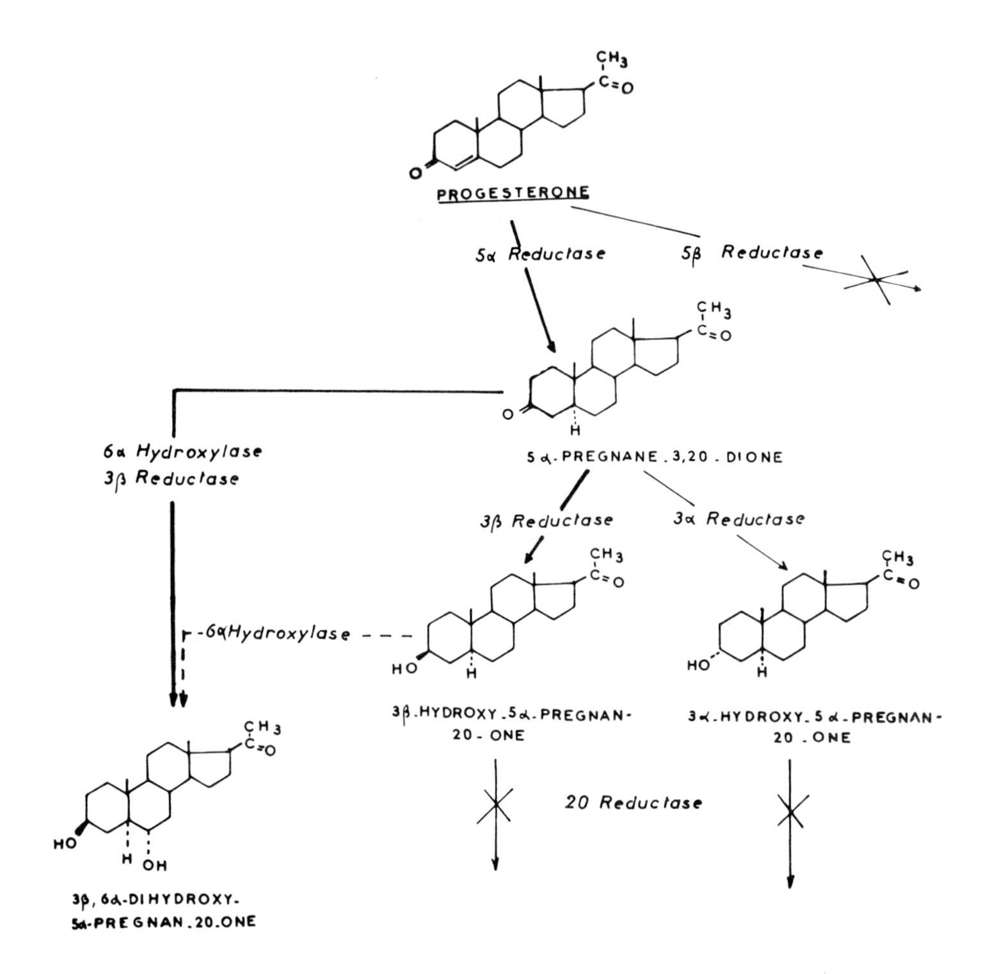

Fig. 9. Metabolic pathway of progesterone incubated with liver cell lines from 20-day-old fetuses and 6-day-old rats. The enzymic route to 3β,6α-dihydroxy-5α-pregnan-20-one is still unknown.

and 6-day-old post-natal rats. The 6α-hydroxylation has only been observed in a few instances[22,77,79] whereas 6β-hydroxylation has been found in numerous tissues including the liver. This hydroxylation in 6α, which is linked to pregnancy, appears as a fetal-liver metabolic route which may be repressed after birth by a hypothalamo-pituitary mechanism which of course does not occur in tissue culture. The sequence of metabolic reactions which lead to 3β,6α-dihydroxy-5α-pregnan-20-one is still not known nor is its regulation. It is hoped that liver cell culture will be a major tool in elucidating this problem.

Metabolism of a xenobiotic compound, safrole, by liver cells in culture[19,88,89]

Establishment of the metabolic map[19]

Safrole is a moderate hepatocarcinogen[80] and a relatively active co-carcinogen[81]. The main metabolites known at present are the 1'-hydroxy-safrole (compound IV in Fig. 11) and the 2',3'-dihydroxysafrole (compound III)[83,84]. The first compound was thought to be ultimate carcinogen. But the epoxide formation is considered, since the work of Grover *et al.*[84], Selkrik *et al.*[85] and Jerina and Daly[86], as one of the most probable if not the principal pathway leading to active carcinogens.

The next step in the epoxide metabolism is the opening of the three--membered ring by the enzyme epoxide hydratase studied by Oesch[87]. This enzyme leads to the formation of a transdihydro-diol which is a way to inactivate the oxide. Research on the metabolism of safrole has been conducted both in the rat and rat-liver cell cultures. Fig. 11 shows the five main metabolic pathways which have been found in the rat and in liver -cell cultures[88,89]:

(1) the epoxidation and the hydroxylation of the allylic chain with, in addition, isomerization of the double bond and its reduction;

(2) the opening of the methylenedioxy ring into eugenol (compound XI) and the loss of the methylene and the methoxy groups;

(3) the association of allylic chain modification and methylenedioxy ring metabolism;

(4) the methylenedioxy ring metabolism and the allylic chain shortening (production of *p*-cresol);

(5) the occurrence in the rat of at least five glucuronides of allyl-catechol (compound XII), propylcatechol, 1'-hydroxysafrole (compound IV), 1'-hydroxyepoxysafrole (compound V) and *p*-cresol.

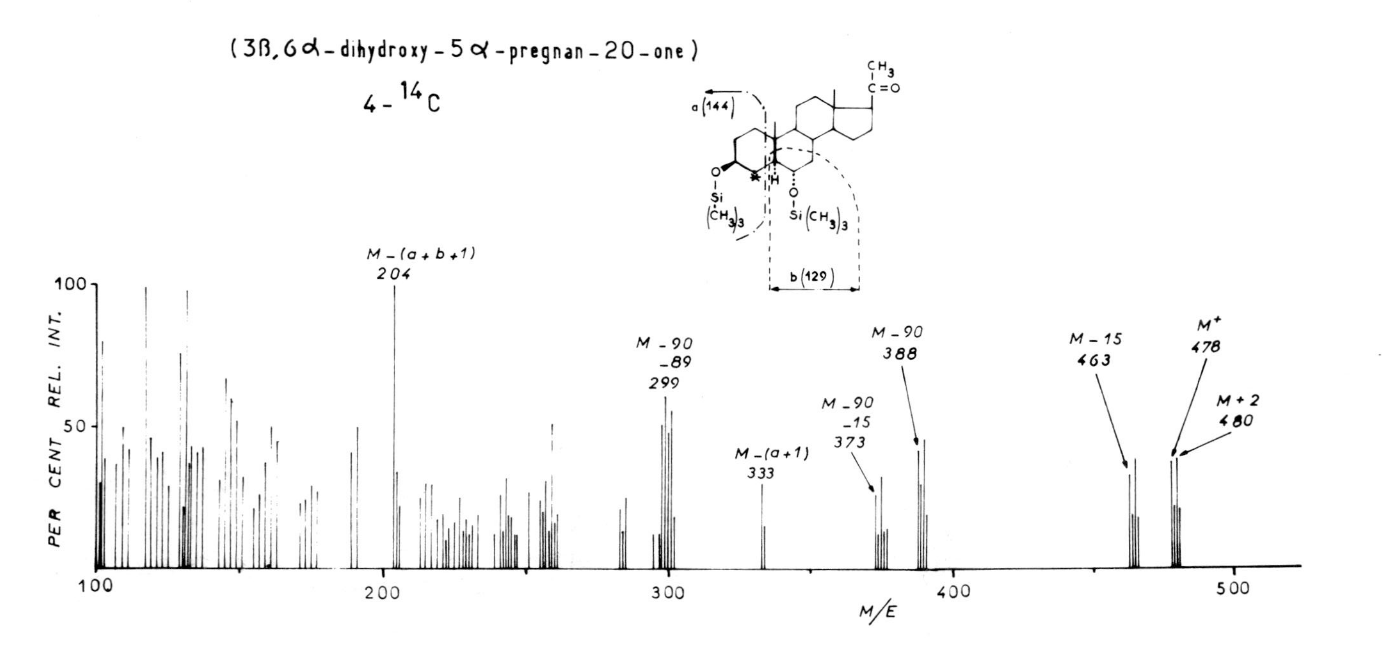

Fig. 10. Mass spectrum of TMS derivative of a progesterone metabolite after incubation of 4-^{14}C-progesterone in the liver cell line and found by radioscanning from TLC pre-purification. As shown on the formula, the occurrence of doublet peaks due to ^{12}C and ^{14}C fragments allowed us to identify easily hydroxylation in the 6 position.

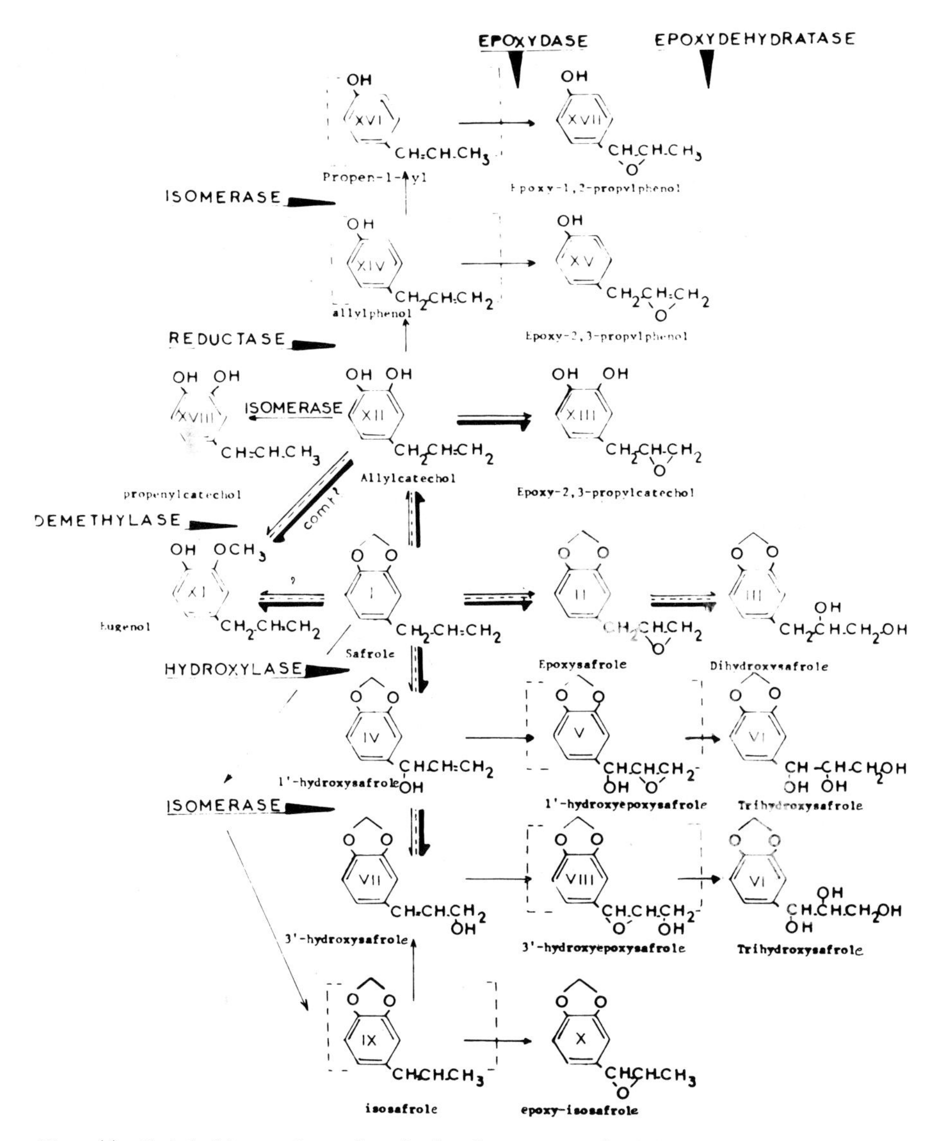

Fig. 11. Metabolic routes of safrole demonstrated either in rat liver homogenate (dotted arrows) or in rat urine (plain arrows) after administration by gastric intubation of one dose of safrole (1.2 g/kg), or in rat-liver cell lines (dark arrows) after 24-h incubation of safrole (1.09 mg/25-cm^2 flask in 5 ml medium). The postulated enzymes are shown. Isosafrole and hydroxy-epoxides between dotted brackets are postulated since precursors and end-products were found. It may be concluded that these hydroxy-epoxides exhibit a high sensitivity to the action of epoxide hydratase, at least in their non-conjugated form.

Action of effectors

The use of GC-MF not only appears as an outstanding tool to elucidate unknown xenobiotic compounds and their metabolism in a quantitative manner but also is the only way to study the action of effectors that must be quantitated such as the induction of enzymes of the endoplasmic reticulum by phenobarbital and by co-cancerogens such as 3-methylcholanthrene. It has been shown that the two main metabolic pathways of safrole, allyl chain oxidation and methylenedioxy ring metabolism, are in competition and that the production of epoxycatechol also arises directly from epoxysafrole (Fig. 12).

Use of liver cell culture

Fig. 11 shows that the liver cell lines produced in the laboratory maintain most of the enzyme by which safrole is metabolized through the two main pathways: side-chain metabolism and methylenedioxy ring opening. The main interests of using liver cell culture for which GC-MS and GC-MF are the only means of study owing to the paucity of tissue available are:

direct estimation of the genuine liver metabolism, because no other tissue interacts before urinary excretion as in the whole animal;

direct compartmental study within the cell;

the use of intermediary compounds that can be produced at will by batch cell cultures from the radioactive safrole;

direct study of the action of effectors; and

direct study of competition between the precursor (*i.e.* safrole) and metabolites (*i.e.* eugenol, estragol) which are also analog compounds.

Fig. 12 shows that incubation of safrole or of 2',3'-epoxysafrole leads to the production of 1'-hydroxysafrole whereas 2',3'-epoxysafrole leads clearly to 2',3'-epoxypropylcatechol. The same experiments in presence of phenobarbital or 3-methylcholanthrene have allowed us to demonstrate a different inducing effect of these compounds on the two types of metabolism[89]. Eugenol (compound XI), when incubated with the cell, undergoes a specific metabolism through the epoxide-diol pathway (Fig. 12). Fig. 13a clearly demonstrates the invaluable use of GC-MS with capillary column to analyze the metabolism of safrole incubated during 48 h (one dose each 24 h) with the ♀155FR30 cell line; and Fig. 13b shows the metabolic events when safrole and eugenol are incubated together for 24 h in the same cell line. Another example of the utmost interest of GC-MF with the cell culture is given by Fig. 14 which shows the effects of

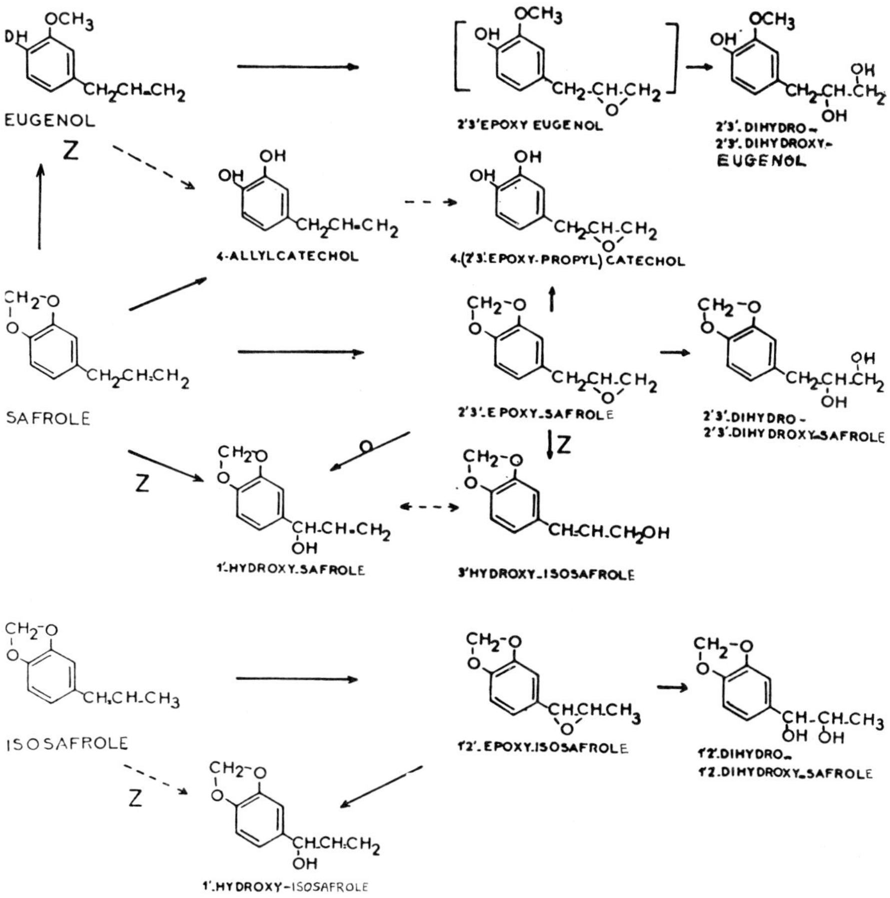

Fig. 12. Metabolisms of safrole, isosafrole, 2',3'-epoxysafrole and eugenol incubated with rat-liver cell lines demonstrating:

(1) interrelations between the epoxy and the monohydroxy compounds since 1'-OH-safrole and 3'-OH-isosafrole arise from both safrole and 2',3'--epoxysafrole, and 1'-OH-isosafrole from isosafrole and 1',2'-epoxyiso-safrole;

(2) interrelations between the epoxidation of the allylic chain and the opening of the methylenedioxy ring in the production of 4(2',3'-epoxy-propyl)catechol;

(3) the metabolism of eugenol as a metabolite of safrole and as a precursor of its own metabolism.

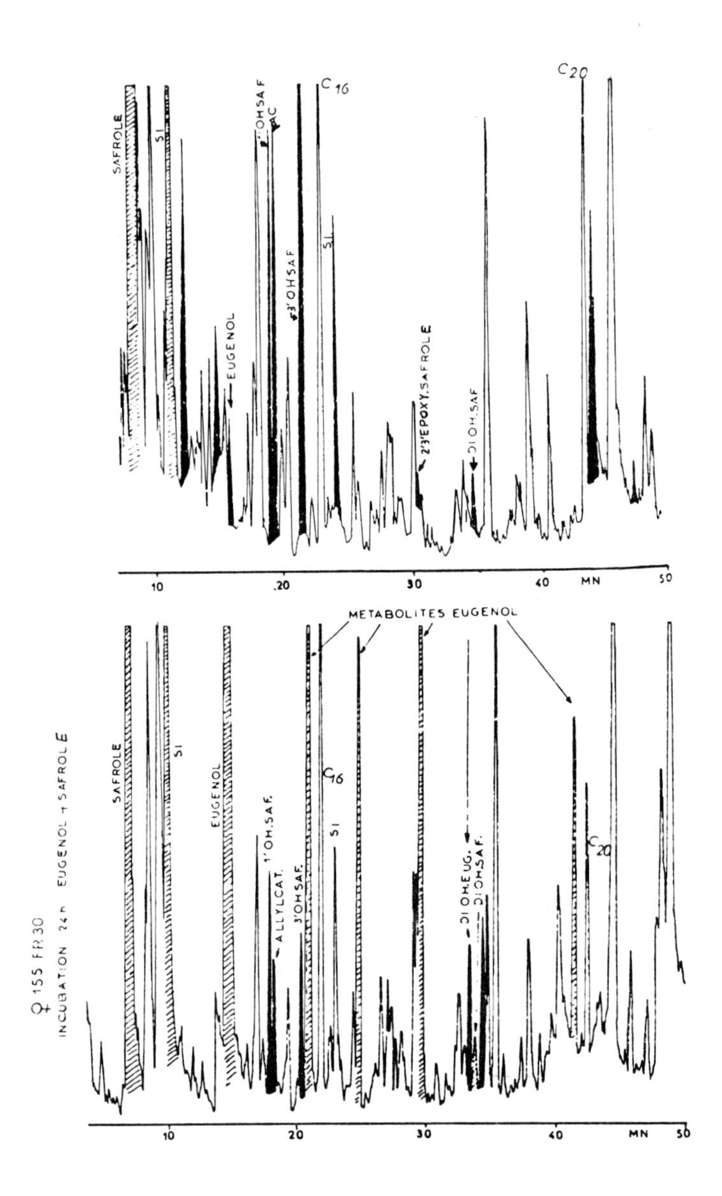

Fig. 13. Capillary gas chromatography of metabolites form safrole and eugenol incubated with a liver cell line at the 30th replication originating from a 155-day-old female adult rat (♀155FR30). Fig. 13a (upper panel) shows the metabolites found after 48-h incubation of two doses of safrole (1.09 mg/24 h). Fig. 13b (lower panel) shows the metabolites found after 24-h incubation of safrole and eugenol together.

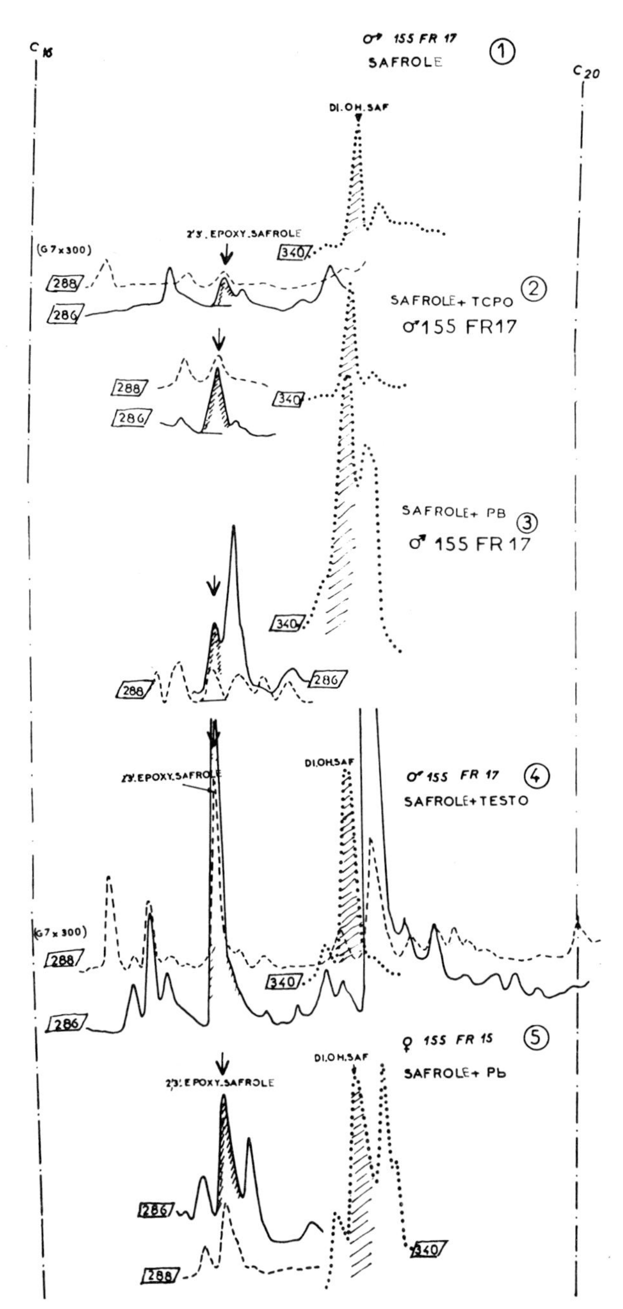

Fig. 14. Study of GC-MF of the action of effectors on liver cell lines incubated with safrole (1.09 mg/25-cm^2 dish in 5 ml of culture medium): male subline (♂155FR17) for Fig. 14-1 to 14-4 and female subline (♀155FR15) for Fig. 14-5. See text for explanations. Epoxides were identified as the Cl-TMS derivatives, either ^{35}Cl-TMS-safrole for *m/e*=286, or ^{37}Cl-TMS-safrole for *m/e*=288.

3,3,3-trichloro-1,2-epoxypropane (TCPO, an inhibitor of epoxydehydratase) and of phenobarbital and testosterone on these liver cell lines incubated with safrole. A comparison of Figs. 14-1 and 14-2 shows the effect of adding TCPO (0.1 µg/25-cm^2 dish in 5 ml medium to the ♂155FR17 cell line: it increases the production of epoxysafrole (*m*/*e*=286 for the molecular ion of ^{35}Cl-TMS and *m*/*e*=288 for ^{37}Cl-TMS derivatives) by partly blocking the epoxide hydratase which produces dihydro-dihydroxysafrole (*m*/*e*=340). The production of epoxysafrole is increased by an order of 5. Compared with Fig. 14-1, Fig. 14-4 demonstrates the important inducing effect of phenobarbital and testosterone on the same cell line on the metabolism of safrole through the epoxide-diol route and the direct effect of testosterone (2.1 µg/25-cm^2 dish in 5 ml of medium). Fig. 14-5 demonstrates that the female liver cell line ♀155FR15 is also inducible by phenobarbital for the production of epoxysafrole and safrole.

CONCLUSION

These examples, we hope, demonstrate clearly that gas-phase analysis methodologies, and especially capillary GC-MS and capillary column-GC-MF, supply an invaluable and an indispensable link to study metabolic pathways not only in cell culture, but also in the animal, since it is an absolute requirement to conduct comparative studies in both biological systems. Enzymic reactions under normal conditions or under the action of effectors are quantitatively and specifically determined by the assay of every precursor and substrate. The use of ^{14}C-compounds in cell culture allows us to establish the interconnections between different pathways and also the competitive action of effectors on these different pathways. For cell biologists and cell cultivators, mass fragmentography has been the only way to use the cell lines produced as an outstanding biological system for quantitation in life sciences.

ACKNOWLEDGEMENTS

This work has been supported during the last five years by grants from "Secrétariat d'État aux Universités" and "l'Université de Dijon" (Vth and VIth Plans), from "Institut National de la Santé et de la Recherche Médicale: Actions Thématiques Programmées": 71.729-331, 74.13.817/AU, 74.14.89 AU/405, "Hormones et cancers": 24.75.47, from "Délégation Générale à la Recherche Scientifique et Technique: Action Concertée

Coordonnée": 72.704.75, from "Fondation pour la Recherche Médicale Française" and from "Ligue Nationale Française Contre le Cancer".

This review paper has highly benefited from the invaluable technical help of Mrs. A. Athias, O. Bonnard, M. Morinière and N. Pitoiset, and Mr. J.P. Morizot.

This article partly summarizes the researches done by F. Barbier--Chapuis, R.J. Bègue, P. Bournot, M. Chessebeuf, M. Delaforge, J. Desgrès, C. Frelin, M. Guiguet, P. Janiaud, P. Lévi, J.C. Lhuguenot, G. Maume, D. Mesnier, J.M. Moalic and M. Prost, as shown by their referenced publications and for which we acknowledge their outstanding collaboration.

REFERENCES

1 J.B. Gurdon, *J. Embryol. Exp. Morphol.*, 10 (1962) 622.
2 P. Rous and F.S. Jones, *J. Exp. Med.*, 23 (1916) 549.
3 A. Moscona, *J. Anat.*, 86 (1952) 287.
4 M.W. Cavanaugh, *J. Exp. Zool.*, 128 (1955) 573.
5 I. Harary and B. Farley, *Science*, 131 (1960) 1674.
6 H. Katsuta and T. Tanaoka, *Jap. J. Exp. Med.*, 35 (1965) 209.
7 L.E. Gerschenson, M. Anderson, J. Molson and J. Okigaki, *Science*, 170 (1970) 859.
8 P. Padieu, C. Lallemant, F. Barbier and M. Chessebeuf, in *VIIth FEBS Meeting, Varna, 1971*, p. 298A.
9 M. Chessebeuf, A. Olsson, P. Bournot, J. Desgrès, M. Guiguet, G. Maume, B.F. Maume, B. Périssel and P. Padieu, *Biochimie*, 56 (1974) 1365.
10 M.J. O'Hare and A.M. Neville, *J. Endocrinol.*, 56 (1973) 537.
11 B.F. Maume and M. Prost, *C.R. Soc. Biol.*, 167 (1973) 1427.
12 B.F. Maume, M. Prost and P. Padieu, in A. Frigerio and N. Castagnoli (Editors), *Advances in Mass Spectrometry in Biochemistry and Medicine*, Spectrum Publications, New York, Vol. 1, 1976, p. 525.
13 J. Paul, *Cell and Tissue Culture*, Livingstone, Edinburgh, London, 1970, p. 430.
14 C.B. Wigley, *Differentiation*, 4 (1975) 25.
15 C. Frelin and P. Padieu, *Biochimie*, (1976), in press.
16 F. Barbier, B.F. Maume and P. Padieu, in A. Frigerio and N. Castagnoli (Editors), *Mass Spectrometry in Biochemistry and Medicine*, Raven Press, New York, 1974, p. 119.
17 J. Desgrès, M. Guiguet, R.J. Bègue and P. Padieu, in A. Frigerio and N. Castagnoli (Editors), *Advances in Mass Spectrometry in Biochemistry and Medicine*, Spectrum Publications, New York, Vol. 1, 1976, p. 139.
18 J. Desgrès, M. Guiguet, R.J. Bègue and P. Padieu, in *Organisation des Laboratoires - Biologie Prospective, 3ème Colloque de Pont à Mousson*, Expansion Scientifique Française, Paris, 1975, p. 471.
19 M. Delaforge, P. Janiaud, M. Chessebeuf, P. Padieu and B.F. Maume, in A. Frigerio (Editor), *Advances in Mass Spectrometry in Biochemistry and Medicine*, Vol. 2, 1976, in press.
20 E.C. Horning, in K.B. Eik-Nes and E.C. Horning (Editors), *Gas Phase Chromatography of Steroids*, Springer-Verlag, Berlin, 1968, p. 3.
21 D. Henneberg, *Fresenius' Z. Anal. Chem.*, 183 (1961) 12.
22 R.J. Bègue, J. Desgrès, J.A. Gustafsson and P. Padieu, *J. Steroid Biochem.*, 7 (1976) 211.

23 R.I. Reed, *Ion Production by Electron Impact*, Academic Press, London, 1962.
24 M.S.B. Munson and F.H. Field, *J. Amer. Chem. Soc.*, 88 (1962) 2621.
25 H.D. Beckey, *Angew. Chem.*, 81 (1969) 662.
26 H.D. Beckey, *Int. J. Mass Spectrom. Ion Phys.*, 2 (1969) 500.
27 E.C. Horning, M.G. Horning, D.I. Carroll, I. Dzidic and N. Stillwell, *Anal. Chem.*, 45 (1973) 936.
28 C.C. Sweeley, W.H. Elliott, I. Fries and R. Ryhage, *Anal. Chem.*, 38 (1966) 1549.
29 L. Sweetman, W.L. Nyhan, P.D. Klein and P.S. Szczepanik, in P.K. Klein and S.V. Peterson (Editors), *Proc. First International Conference on Stable Isotopes in Chemistry, Biology and Medicine*, U.S. Department of Commerce, 1973, p. 404.
30 P. Bournot, M. Chessebeuf, G. Maume, A. Olsson, B.F. Maume and P. Padieu, in A. Frigerio and N. Castagnoli (Editors), *Mass Spectrometry in Biochemistry and Medicine*, Raven Press, New York, 1974, p. 151.
31 B.F. Maume, M. Prost and P. Padieu, in A. Frigerio and N. Castagnoli (Editors), *Advances in Mass Spectrometry in Biochemistry and Medicine*, Spectrum Publications, New York, Vol. 1, 1976, p. 525.
32 M. Prost, D. Mesnier, P. Bournot, J.F. Jeannin and B.F. Maume, in *Organisation des Laboratoires - Biologie Prospective, 3ème Colloque de Pont à Mousson*, Expansion Scientifique Française, 1975, p. 317.
33 C.G. Hammar and R. Hessling, *Anal. Chem.*, 43 (1971) 298.
34 C.G. Hammar and R. Hessling, *Anal. Chem.*, 25 (1968) 532.
35 V.L. Tal'Rose, V.A. Pavlenko, G.D. Tantsyrev, V.D. Grishin, L.N. Ozerov, A.E. Rafal'Don and M.D. Shutov, *Prib. Tekn. Eksper.*, 10 (1965) 130.
36 C.J.W. Brooks, A.R. Thawley, P. Rocher, B.S. Middleditch, G.M. Anthony and N.G. Stillwell, *J. Chromatogr. Sci.*, 9 (1971) 35.
37 T.A. Baillie, C.J.W. Brooks and B.S. Middleditch, *Anal. Chem.*, 44 (1972) 30.
38 S.H. Koslow, F. Cattabeni and E. Costa, *Science*, 176 (1972) 177.
39 T.E. Gaffney, C.G. Hammar, B. Holmstedt and R.E. McMahon, *Anal. Chem.*, 43 (1971) 307.
40 R.A. Hites and K. Biemann, *Anal. Chem.*, 42 (1970) 855.
41 C.C. Sweeley, B.D. Ray, W.I. Wood, J.F. Holland and M.I. Krichevsky, *Anal. Chem.*, 42 (1970) 1505.
42 R. Reimendal and J. Sjövall, *Anal. Chem.*, 44 (1972) 21.
43 J.F. Holland, C.C. Sweeley, R.E. Thrush, R.E. Teets and M.A. Bieber, *Anal. Chem.*, 45 (1973) 308.
44 L. Baczynskyj, D.J. Duchamp, J.F. Zieseri, Jr., and U. Axen, *Anal. Chem.*, 45 (1973) 479.
45 W.F. Holmes, W.H. Holland, B.L. Shore, D.M. Bier and W.R. Sherman, *Anal. Chem.*, 45 (1973) 2063.
46 J.T. Watson, D.R. Pelster, B.J. Sweetman, J.C. Frolich and J.A. Oates, *Anal. Chem.*, 45 (1973) 2063.
47 R.M. Caprioli, W.F. Fies and M.S. Tory, *Anal. Chem.*, 46 (1974) 453A.
48 D. Young, J.F. Holland, J.N. Gerber and C.C. Sweeley, *Anal. Chem.*, 47 (1975) 2373.
49 B. Samuelson, M. Hamberg and C.C. Sweeley, *Anal. Biochem.*, 38 (1970) 301.
50 U. Axen, K. Green, D. Hörlin and B. Samuelson, *Biochem. Biophys. Res. Commun.*, 45 (1971) 519.
51 M.A. Bieber, C.C. Sweeley, D.J. Faulkner and M.R. Petersen, *Anal. Biochem.*, 47 (1972) 264.
52 L. Bertilsson, A.J. Atkinson, Jr., J.R. Althaus, A. Härfast, J.E. Lindgren and B. Holmstedt, *Anal. Chem.*, 44 (1972) 1434.

53 B.F. Maume, P. Bournot, J.C. Lhuguenot, C. Baron, F. Barbier, G. Maume, M. Prost and P. Padieu, *Anal. Chem.*, 45 (1973) 1073.
54 M. Prost and B.F. Maume, in A. Frigerio and N. Castagnoli (Editors), *Mass Spectrometry in Biochemistry and Medicine*, Raven Press, New York, 1974, p. 139.
55 M. Prost and B.F. Maume, *J. Steroid Biochem.*, 5 (1974) 133.
56 P. Bournot, B.F. Maume and P. Padieu, *Biomed. Mass Spectrom.*, 1 (1974) 29.
57 P. Bournot, M. Prost and B.F. Maume, *J. Chromatogr.*, 112 (1975) 617.
58 M. Chessebeuf, A. Olsson, P. Bournot, J. Desgrès, M. Guiguet, G. Maume, B.F. Maume, B. Périssel and P. Padieu, in L. Gerschenson and B. Thompson (Editors), *Gene Expression and Carcinogenesis in Cultured Liver*, Academic Press, New York, 1974, p. 94.
59 B.F. Maume and P. Padieu, *Journées de Pharmacologie Moléculaire, Université Paul Sabatier, Toulouse, 28-29 avril 1976.*
60 C. Fenselau, *Appl. Spectrosc.*, 28 (1974) 305.
61 A.M. Lawson, *Clin. Chem.*, 21 (1975) 803.
62 A. Frigerio and N. Castagnoli (Editors), *Mass Spectrometry in Biochemistry and Medicine*, Raven Press, New York, 1974.
63 A. Frigerio and N. Castagnoli (Editors), *Advances in Mass Spectrometry in Biochemistry and Medicine*, Spectrum Publications, New York, Vol. 1, 1976.
64 A. Frigerio (Editor), *Proceedings of the International Symposium on Gas Chromatography-Mass Spectrometry*, Tamburini Publisher, Milan, 1972.
65 A. Frigerio (Editor), *Advances in Mass Spectrometry in Biochemistry and Medicine*, Spectrum Publications, New York, Vol. 2, 1976, in press.
66 A.P. De Leenheer and R.R. Roncucci (Editors), *Quantitative Mass Spectrometry in Life Sciences*, Elsevier, Amsterdam, Oxford, New York 1977.
67 R.W. Kelly, *J. Chromatogr.*, 54 (1971) 345.
68 C.H. Draffan, R.A. Clare and F.M. Williams, *J. Chromatogr.*, 75 (1973) 145.
69 B.F. Maume, P. Bournot, J. Durand, J.C. Lhuguenot, G. Maume, M. Prost and P. Padieu, *Organisation des Laboratoires - Biologie Prospective, 2ème Colloque de Pont à Mousson*, Expansion Scientifique Française, Paris, 1972, p. 637.
70 J.A. McCloskey, R.N. Stillwell and A.M. Lawson, *Anal. Chem.*, 40 (1968) 233.
71 P. Bournot, 1973, personal communication.
72 V. Buanossisi, G. Sato and A.I. Cohen, *Proc. Nat. Acad. Sci. U.S.*, 48 (1962) 1148.
73 R.W. Pierson, *Endocrinology*, 81 (1967) 693.
74 G. Maume, 1976, in preparation.
75 H.F. Acevedo, H.S. Stricker, J. Gilmore, B.A. Vela, A.E. Campbell and B.J. Arras, *Amer. J. Obstet. Gynecol.*, 102 (1968) 867.
76 R.J. Bègue, J. Desgrès, J.A. Gustafsson and P. Padieu, *J. Chromatogr. Sci.*, 12 (1974) 763.
77 C.H.L. Schackleton, J.A. Gustafsson and J. Sjövall, *Steroids*, 17 (1971) 265.
78 P. Eneroth, M. Ferngren, J.A. Gustafsson, B. Ivemarck and A. Stenberg, *Acta Endocrinol.*, 70 (1972) 113.
79 K. Fotherby, *Advan. Biosci.*, 3 (1968) 43.
80 E.L. Long, W.M. Hansen and A.A. Nelson, *Fed. Proc., Fed. Amer. Soc. Exp. Biol.*, 20 (1961) 287.
81 F.J. McPherson, J.W. Bridges and D.V. Parke, *Biochem. Pharmacol.*, 25 (1976) 1345.
82 P.G. Wislocki, P. Borchert, E.C. Miller and J.A. Miller, *Proc. Amer. Cancer Soc.*, 13 (1972) 12.

83 W.G. Stillwell, M.J. Carman, L. Bell and M.G. Horning, *Drug Metab. Disposition*, 2 (1974) 489.
84 P.L. Grover, A. Hewer and P. Sims, *FEBS Lett.*, 18 (1971) 76.
85 J.K. Selkrik, E. Huberman and C. Heidelberger, *Biochem. Biophys. Res. Commun.*, 43 (1971) 1010.
86 D.M. Jerina and J.W. Daly, *Science*, 185 (1976) 79.
87 F. Oesch, *J. Biol. Chem.*, 25 (1976) 79.
88 P. Janiaud, M. Delaforge, Ph. Levi, B.F. Maume and P. Padieu, *Soc. Biol., Dijon, 17 juin 1976.*
89 P. Janiaud, M. Delaforge, Ph. Levi, B.F. Maume and P. Padieu, in *Coll. Int. CNRS sur la Cancérogénèse Chimique, Menton, 4-9 juillet 1976.*

QUANTITATIVE MASS FRAGMENTOGRAPHIC DETERMINATION OF UNLABELED AND DEUTERIUM-LABELED PROPOXYPHENE IN PLASMA

MEASUREMENT OF STEADY-STATE PHARMACOKINETICS IN DOGS

H.R. SULLIVAN and R.E. McMAHON

The Lilly Research Laboratories, Indianapolis, Ind. 46206 (U.S.A.)

SUMMARY

A quantitative mass fragmentographic method for the determination of plasma levels of d-propoxyphene has been developed. In this method, d-propoxyphene-benzyl-d_7 served as the internal standard. The ratio of d-propoxyphene-d_0 to d-propoxyphene-d_7 was calculated from the *m/e* 208 *vs.* *m/e* 215 ion intensity ratio determined by gas chromatography-mass spectrometry in the electron impact mode. This method was applied to the determination of plasma concentrations of d-propoxyphene in single dose studies in animals. By the use of a second label, d-propoxyphene-benzyl-d_2, studies have now been initiated in which an attempt was made to investigate drug pharmacokinetics in the steady state. The benzyl moiety was selected as the labeling position in the d-propoxyphene to be used since this position is not involved in metabolism and therefore should not lead to undesirable isotope effects. This expectation was confirmed by dosing a dog with a mixture of d-propoxyphene-d_0 and d-propoxyphene-d_2 and determining the propoxyphene-d_2/propoxyphene-d_0 ratio in plasma at various times. Since this d_2/d_0 ratio in plasma was the same as that in the administered dose and since it did not change over the time period used (0-8 h) it is evident that the two compounds were pharmacologically equivalent in the body.

In a typical steady-state experiment a single daily dose of d-propoxyphene (5 mg/kg) was administered orally to dogs for 19 consecutive days. On day 20 the daily dose of the drug was substituted with a pulse dose of 5 mg of propoxyphene-d_2 per kg. Total plasma levels of propoxyphene as well as the propoxyphene-d_2/propoxyphene-d_0 ratio were then monitored for 48 h by quantitative mass fragmentography. During the first hour, levels of propoxyphene-d_0 increased, indicating some rapid exchange of propoxyphene-d_2 with a readily available bound pool of propoxyphene-d_0. However, the extent of this exchange was not great. Rather, the ratio of propoxyphene-d_2/propoxyphene-d_0 continued to change with time and did not reach equilibrium even at 48 h. These results show the existence of "deep" pools of tissue-bound propoxyphene in chronically dosed dogs. The equilibration of bound drug in these deep pools with a pulse dose is clearly a slow process.

These results have several implications. For example, tissue disposition studies in chronically dosed animals which are commonly performed by pulse dosing with ^{14}C-labeled drug are probably not reliable measures of drug disposition during chronic dosing.

INTRODUCTION

Since its first description in 1953[1], d-propoxyphene has become one of the most widely used analgesics in current medical practice. In 1968 Wolen and Gruber[2] described a gas-liquid chromatographic (GLC) method for the quantitative determination of plasma levels of d-propoxyphene and, in addition, reported the first pharmacokinetic study of this drug in man. This GLC method lacked sufficient sensitivity at low drug levels and therefore was of limited utility for the study of the pharmacokinetics of d-propoxyphene in laboratory animals. Indeed, a more sensitive and specific method was needed for more sophisticated pharmacokinetic studies in both animals and in man.

The technique of quantitative mass fragmentography (QMF)[3,4], in which the mass spectrometer is used as a highly specific and sensitive detector for GLC has proven to be of great value in drug research. Recently, QMF methods for the quantitation of d-propoxyphene in body fluids have been developed and have been applied to single-dose studies in laboratory animals[5] and in man[6]. Results obtained from these studies[5,6] suggested that substantial pools of tissue-bound drug existed and that these pools influenced plasma kinetics to an appreciable extent. Further investigation of these problems necessitated a modification of the QMF method in such a manner as to enable the simultaneous monitoring of multiple drug pools. In preliminary studies this modification has been satisfactorily accomplished by the utilization of "pulse" dosing with deuterium-labeled drug whose molecular weight differed from that of the deuterium-labeled drug used as internal standard for QMF measurements. A similar method had been developed for the investigation of multi-dose kinetics of methadone in man[7]. The development and subsequent application of this QMF method is the subject of this communication.

DEVELOPMENT OF METHOD

The recently developed QMF method[5] using gas chromatography-mass spectrometry in the electron impact mode (GC-MS-EI) in conjunction with deuterium labeling has been successfully applied to the determination of plasma concentrations of d-propoxyphene-d_0 (Fig. 1) in laboratory animals for 24 h after oral administration of a single dose of 20 mg of the drug per kg. The internal standard selected for this method was d-propoxyphene--benzyl-d_7 (Fig. 1). Kinetic data obtained from single-dose studies, however, may not represent kinetics in effect at the steady state because of possible changes in distribution and elimination characteristics upon multi-dosing. Study of d-propoxyphene kinetics at the steady state, therefore, necessitated modification of this QMF method to permit the co--determination of labeled and unlabeled d-propoxyphene in plasma after administration of a single "pulse" or labeled dose during a multi-dose study. By the use of a second labeled drug, d-propoxyphene-benzyl-d_2 (Fig. 1) to be used as the pulse dose, studies have been initiated in which an attempt was made to investigate d-propoxyphene pharmacokinetics at the steady state condition.

Mass fragmentography

An LKB-9000 combined gas chromatograph-mass spectrometer fitted with an accelerating voltage alternator (AVA) accessory unit[4] was used for QMF measurements. Separations were accomplished with a 1.2-m silanized glass column packed with 1% UV-W98 methylvinyl silicone gum rubber on 80-100 mesh Gas-Chrom Q. The column temperature was maintained at 200° while the ionization potential and trap current were 22 eV and 60 μA, respectively.

The mass fragmentation pattern obtained by GC-MS analysis of a mixture of d-propoxyphene-d_0 and internal standard, d-propoxyphene-d_7, is shown in Fig. 2. The ion, *m/e* 208 (M-131), was selected as the basis of the QMF method and resulted from the co-elimination of the propionoxy and

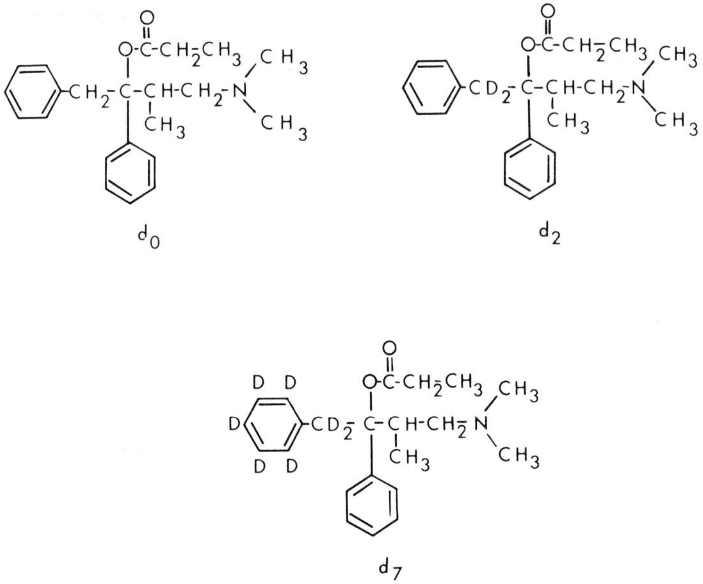

O
‖
O-C-CH_2CH_3 CH_3
CH_2-C-CH-CH_2-N
CH_3 CH_3

d_0

O
‖
O-C-CH_2CH_3 CH_3
CD_2-C-CH-CH_2-N
CH_3 CH_3

d_2

D D
O
‖
O-C-CH_2CH_3 CH_3
D CD_2-C-CH-CH_2-N
D D CH_3 CH_3

d_7

Fig. 1. Structures of d-propoxyphene-d_0, d-propoxyphene-d_2 and internal standard, d-propoxyphene-benzyl-d_7, used in this study.

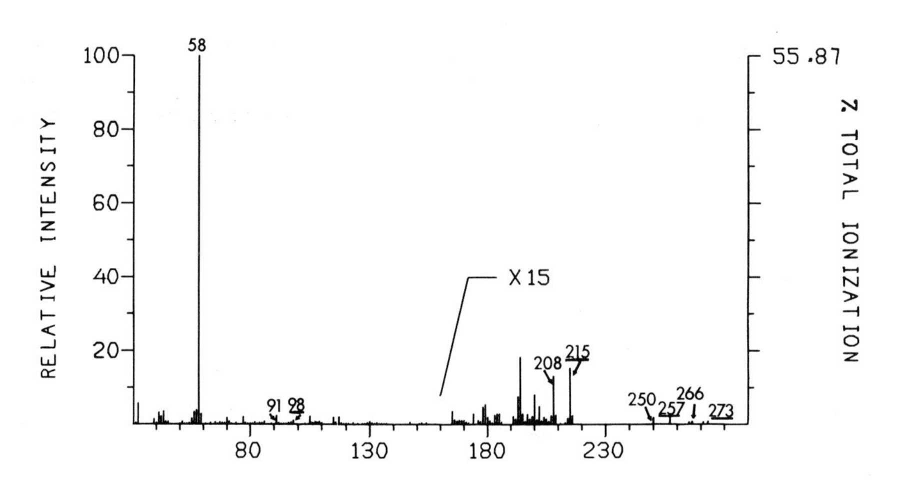

Fig. 2. Mass fragmentation (EI) pattern of a mixture of d-propoxyphene-d_0 and d-propoxyphene-d_7.

dimethylaminomethyl moieties from M^{+}. This ion was a relatively weak ion (5% of base peak) and therefore limited the sensitivity of the method. There was not, however, a more abundant ion present that could be conveniently labeled with deuterium. The corresponding M-131 ion obtained from the internal standard appeared at *m/e* 215. Thus, by determining the *m/e* 208 *vs.* *m/e* 215 ion intensity ratio by QMF analysis the ratio of d-propoxyphene-d_0 to internal standard in a sample could be calculated. For the present modification the second labeled drug, d-propoxyphene-d_2, to be used as the "pulse" dose has a corresponding M-131 ion at *m/e* 210. The co-determination of *m/e* 208 *vs.* *m/e* 215 and *m/e* 208 *vs.* *m/e* 210 ion intensity ratios for a particular sample would, therefore, permit the calculation of the concentration of d-propoxyphene-d_0 and d-propoxyphene-d_2 present in a sample.

Standard curves

Standard curves in which the observed peak height ratios were plotted against known, molar ratios, determined gravimetrically, were constructed and used in this method. Twelve 4-ml solutions of dog plasma containing d-propoxyphene-d_0 and d-propoxyphene-d_2 in concentrations ranging from 14.8 pmole/ml to 2.21 nmole/ml and including two blanks were prepared. Internal standard, d-propoxyphene-d_7, was added to each sample, less one blank, at a concentration of 705 pmole/ml. After samples had been allowed to equilibrate for 1 h at room temperature, each sample was adjusted to pH 9.5 with 1 *M* NaOH and was then extracted twice with 4.0-ml portions of distilled *n*-butyl chloride. The combined organic phase from each sample was evaporated to dryness under a stream of nitrogen. The residual material was dissolved in 0.1 ml of toluene for QMF analysis. Initially, the ratio of *m/e* 208 *vs.* *m/e* 215 ion intensities was determined for each sample. This observed peak-height ratio was plotted against the known molar ratio of d-propoxyphene-d_0 to internal standard added to the plasma solution. The standard curve constructed from data obtained from analysis of these plasma extracts was linear throughout the range of concentrations of d-propoxyphene-d_0. With the completion of these analyses the mass spectrometer was refocused and the ratio of *m/e* 210 and *m/e 215* ion intensities was determined for each of these same samples. This observed peak-height ratio was plotted against the known molar ratio of d-propoxyphene-d_2 to internal standard added to the plasma solution. Again, the standard curve constructed from data obtained from this analysis was linear throughout the range of concentrations of d-propoxyphene-d_2.

These standard curves proved to be quite flexible. One important advantage of this type of curve is that the amount of internal standard added can be varied depending upon anticipated sample concentration without requiring construction of additional standard curves. The observed peak-height ratios varied substantially from the actual molar ratios of compounds present in the samples. The mean difference between molar ratios and peak-height ratios for d-propoxyphene-d_0 was 15% whereas that for d-propoxyphene-d_2 was 6%. This difference may well be due, at least in part, to differences in extraction efficiencies of labeled and unlabeled drug and internal standard from plasma. Another possible factor influencing peak-height ratios would be an effect of deuterium isotope on the yield of the M-131 ion from the d-propoxyphene-d_2 and internal standard. Each of these possible sources of error would be corrected for by use of these standard curves. A possible source of error not corrected for could arise from the possible dependence of observed peak-height ratios on the absolute quantity of sample introduced into the ion source of the mass spectrometer. This possibility was investigated, and no differences were observed throughout a range of sample sizes likely to be encountered in the projected study. The time of equilibration of internal standard with plasma was also investigated and no significant difference between one and eight hours was observed.

Accuracy, precision and sensitivity

Solutions of 50 ml each of d-propoxyphene-d_0 at 11.8, 59.0, 295.0, and 1180 pmole/ml; d-propoxyphene-d_2 at 11.7, 58.7, 293.3 and 1173 pmole/ml and mixtures of d-propoxyphene-d_0 and d-propoxyphene-d_2 at 59.0 *vs.* 117.3 and 236.0 *vs.* 176.0 pmole/ml were prepared in dog plasma. After each solution had been allowed to equilibrate at room temperature for 1 h, multiple 4.0-ml aliquots of each solution, along with two 4.0-ml aliquots of control plasma, were taken for QMF analysis. Internal standard, d-propoxyphene-d_7, at a selected concentration was added to each set of samples and to one control sample per set. Samples were extracted with *n*-butyl chloride as previously described, and the combined organic phase from each sample was evaporated to dryness under nitrogen. The residual material from each sample extract was dissolved in 0.1 ml toluene for analysis.

Results obtained from the QMF analysis of these samples are summarized in Table I. These data show this exceptionally precise and accurate method to be more than satisfactory for the co-determination of d-propoxyphene-d_0 and d-propoxyphene-d_2 in a plasma sample. In the

Table I

Analysis of dog plasma samples containing known concentrations of d-propoxyphene-d_0 and/or d-propoxyphene-d_2. Accuracy, precision and sensitivity of QMF method

Drug	Amount of drug added (pmole/ml)	Internal standard added (pmole/ml)	Amount of drug found (pmole/ml ± S.D.)$^{+}$
Propoxyphene	1180.0	723.5	1174.0 ± 6.0
	295.0	723.5	298.0 ± 2.9
	59.0	144.5	58.4 ± 1.5
	11.8	28.9	12.2 ± 0.4
Propoxyphene-d_2	1173.0	723.5	1198.0 ± 8.0
	293.3	723.5	291.5 ± 3.8
	58.7	144.5	58.1 ± 1.2
	11.7	28.9	12.1 ± 0.2
Propoxyphene +	63.1	289.0	64.9 ± 0.9
propoxyphene-d_2	113.2	289.0	111.1 ± 0.9
Propoxyphene +	242.5	289.0	244.8 ± 1.2
propoxyphene-d_2	169.5	289.0	170.7 ± 2.6

$^{+}$ $n = 6$.

conditions described herein, the lower limit of sensitivity for the co-determination of the labeled and unlabeled drug in plasma was about 8 pmole/ml with a coefficient of variation of less than 3.5%. The lower limit of sensitivity for the GLC-FID method for the determination of plasma concentrations of propoxyphene is 74 pmole/ml[8]. The value of the QMF method, over and above increased sensitivity lies, however, in increased precision and its capability of measuring labeled and unlabeled drug simultaneously. The sensitivity of this method is limited only by the abudance of the M-131 ion used in the analysis. Methadone is another example in which the QMF method is similarly limited by ion yield[7]. The QMF sensitivity for methadone is about 16 pmole/ml compared with 50 pmole/ml for the best GLC-FID method. On the other hand, with compounds such as the cannabinoids, which have much more intense ions in their EI fragmentation patterns, sensitivities of 3 pmole/ml or less are not uncommon[9,10].

Labeled and unlabeled d-propoxyphene metabolic equivalency. Absence of isotope effects

The benzyl moiety of d-propoxyphene was selected as the labeling position since previous metabolism studies[10] have shown this position not to be involved in its metabolism and therefore it would not lead to undesirable isotope effects. By the application of this modified QMF method for the simultaneous determination of plasma concentrations of labeled and unlabeled drug, this postulation could now be unambiguously confirmed or denied. A solution of a mixture of approximately 5 mg/ml each of d-propoxyphene-d_0 and d-propoxyphene-d_2 in water was prepared. QMF analysis of an aliquot of this solution revealed the ratio of d-propoxyphene-d_2 (*m/e* 210) *vs.* d-propoxyphene-d_0 (*m/e* 208) in this solution to be 0.82. Using this same solution, a single dose of approximately 5 mg/kg each of labeled and unlabeled d-propoxyphene was administered orally to a fasted dog. Blood samples (10 ml) were drawn from the dog immediately before and at 0.5, 1, 2, 4, 8 and 24 hours after administration of the drug. Plasma fractions were immediately separated by centrifugation at 1000 *g*. Internal standard was added to each plasma sample and the mixture allowed to equilibrate for 1 h at room temperature. Each sample was then processed for QMF analysis by the method previously described.

The purpose of the QMF analysis was to determine not only concentrations of total d-propoxyphene ($d_0 + d_2$) but also to determine the relative concentration of labeled and unlabeled d-propoxyphene in each sample. Each sample extract was analyzed, first, for levels of d-propoxyphene-d_0, determining the *m/e* 208 *vs.* *m/e* 215 ion intensity ratio, and second, for the ratio of concentration of d-propoxyphene-d_2 *vs.* d-propoxyphene-d_0, determining the *m/e* 210 *vs.* *m/e* 208 ion intensity ratio. The data obtained from these analyses were used to calculate the concentration of total propoxyphene ($d_0 + d_2$) present in each sample. Results obtained from these analyses, summarized in Fig. 3, show, in addition the ratio of labeled to unlabeled d-propoxyphene (*m/e* 210 *vs.* *m/e* 208) in every plasma sample to be essentially the same as that of the administered solution (0.82). The lack of significant deviation of this d_2/d_0 ratio from that of the administered solution at any time, therefore, negates any possibility of an isotope effect in the distribution and metabolism of d-propoxyphene-d_2 in dogs.

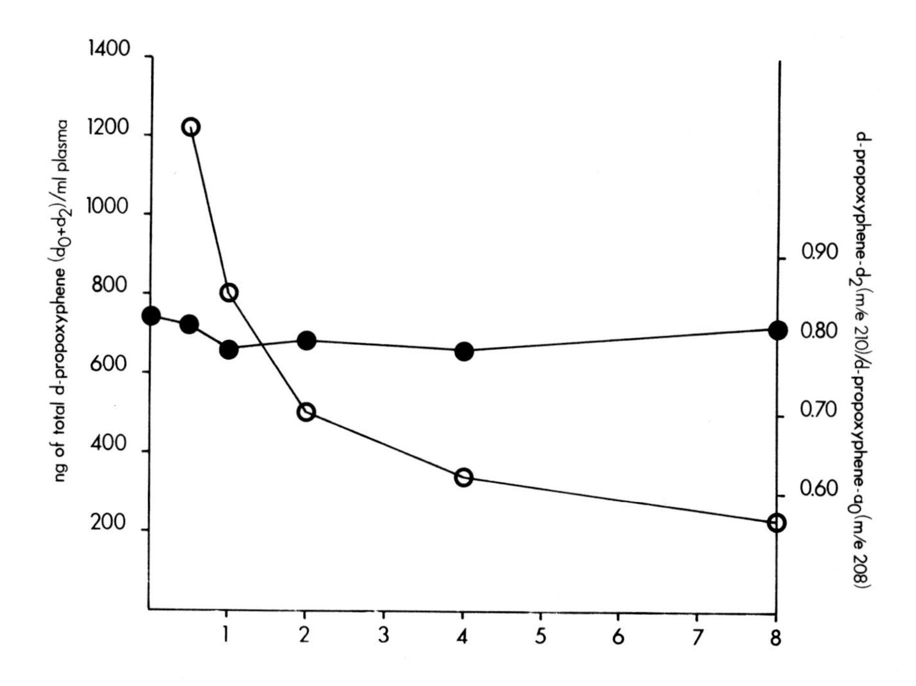

Fig. 3. Plasma concentrations of total propoxyphene (d_0 + d_2) in a dog after oral administration of a single dose, at 5 mg/kg each, of d-propoxyphene-d_0 and d-propoxyphene-d_2 (d_2/d_0 = 0.82).

Pharmacokinetic study of steady state. Existence of "deep" drug pools

The modified QMF assay method described above is now being applied to the investigation of the pharmacokinetics of propoxyphene in the steady state in dogs. In a typical study a single daily dose of d-propoxyphene-d_0 (5 mg/kg) was administered orally to two dogs for 19 consecutive days. On day 20 the daily dose of the drug was substituted with a "pulse" dose of d-propoxyphene-d_2 at 5 mg/kg. Blood samples were drawn 2 h, immediately before and at 0.25, 0.5, 1, 2, 6, 10, 22, 26 and 48 h after administration of the pulse dose. Plasma fractions were separated and internal standard, d-propoxyphene-d_7 (150 ng/ml), was added to each fraction. Plasma samples were processed for QMF analysis by the previously described procedure. Each plasma extract was analyzed for concentrations of d-propoxyphene-d_0, *m/e* 208 *vs.* *m/e* 215, and for the d-propoxyphene-d_2 *vs.* d-propoxyphene-d_0 concentration ratio, *m/e* 210 *vs.* *m/e* 208.

Plasma levels of d-propoxyphene-d_0 and d-propoxyphene-d_2 found in these dogs are shown in Tables II and III. The primary purpose of this

Table II

Relative rates of disposition of d-propoxyphene-d_0 and d-propoxyphene-d_2 in the dog

A female dog, No. 7141, received a single oral dose of d-propoxyphene-d_0 (5 mg/kg) daily for 19 days. On day 20 an oral dose, at 5 mg/kg, of d-propoxyphene-d_2 was substituted for the daily dose of d-propoxyphene-d_0.

Time (h)	d-propoxyphene-d_2/ d-propoxyphene-d_0	d-propoxyphene-d_0 level (ng/ml)	d-propoxyphene-d_2 level (ng/ml)
-2	-	16.2	-
0	-	13.5	-
0.25	19.23	27.5	529
0.50	19.28	24.2	467
1	15.04	19.5	293
2	5.06	26.1	132
6	2.34	28.2	66.0
10	2.51	13.8	34.7
22	0.38	18.6	7.1
26	0.48	13.5	6.5
48	0.31	14.3	4.4

Table III

Relative rates of disposition of d-propoxyphene-d_0 and d-propoxyphene-d_2 in the dog

A female dog, No. 6637, received a single oral dose of d-propoxyphene-d_0 (5 mg/kg) daily for 19 days. On day 20 an oral dose, at 5 mg/kg, of d-propoxyphene-d_2 was substituted for the daily oral dose of d-propoxy-phene-d_0.

Time (h)	d-propoxyphene-d_2/ d-propoxyphene-d_0	d-propoxyphene-d_0 level (ng/ml)	d-propoxyphene-d_2 level (ng/ml)
-2	-	11.9	-
0	-	11.9	-
0.25	6.37	12.8	81.8
0.50	4.77	19.5	93.0
1	6.00	15.0	90.0
2	4.71	10.2	48.0
6	1.18	13.8	16.2
10	0.80	13.1	10.5
22	0.42	11.4	4.8
26	0.36	13.5	4.8
28	0.23	11.1	2.6

study was to demonstrate the applicability of the method, but these results also revealed a number of new and interesting features in the pharmacokinetics of d-propoxyphene. During the first hour after administration of the "pulse" dose, levels of d-propoxyphene-d_0 increased indicating some rapid exchange of d-propoxyphene-d_2 with a readily available pool of unlabeled drug. The extent of this exchange was, however, not great. Rather, the ratio of d-propoxyphene-d_2/d-propoxyphene-d_0 continued to change with time and did not reach equilibrium even at 48 h. These results suggest the existence of "deep" pools of tissue-bound d-propoxyphene in these chronically dosed dogs. The equilibration of bound drug in these deep pools with a pulse dose is clearly a slow process. These results have several implications. For example, tissue disposition studies in chronically dosed animals which are commonly performed by pulse dosing with ^{14}C-labeled drug are probably not reliable measures of drug disposition during chronic treatment.

In summary, the QMF analytical method described here should facilitate the study of steady state pharmacokinetics of drugs in general and, in particular, where multiple drug pools are expected.

REFERENCES

1 A. Pohland and H.R. Sullivan, *J. Amer. Chem. Soc.*, 75 (1953) 4458.
2 R.L. Wolen and C.M. Gruber, Jr., *J. Anal. Chem.*, 40 (1968) 1243.
3 C.-G. Hammar, B. Holmstedt and R. Ryhage, *Anal. Biochem.*, 25 (1968) 532.
4 B. Holmstedt and L. Palmer, in E. Costa (Editor), *Advances in Psychopharmacology*, Raven Press, New York, 1973, p.1.
5 H.R. Sullivan, J.L. Emmerson, F.J. Marshall, P.G. Wood and R.E. McMahon, *Drug Metab. Dispos.*, 2 (1974) 526.
6 R.L. Wolen, E.A. Ziege and C.M. Gruber, Jr., *Clin. Pharmacol. Ther.*, 17 (1975) 15.
7 H.R. Sullivan, F.J. Marshall, R.E. McMahon, L.-M. Gunne, J.H. Holmstrand and E. Änggard, *Biomed. Mass Spec.*, 2 (1975) 197.
8 J.F. Nash, I.F. Bennett, R.J. Bopp, M.K. Brunson and H.R. Sullivan, *J. Pharm. Sci.*, 65 (1975) 429.
9 S. Agurell, B. Gustafsson, B. Holmstedt, K. Leander, J.-E. Lindgren, I. Nilsson, F. Sandberg and M. Asberg, *J. Pharm. Pharmacol.*, 25 (1973) 554.
10 H.R. Sullivan, D.L.K. Kau and P.G. Wood, *Proceedings 2nd International Congress Stable Isotopes*, 1976, in press.
11 S.L. Due, H.R. Sullivan and R.E. McMahon, *Biomed. Mass Spec.*, in press.

MASS FRAGMENTOGRAPHY DETERMINATION OF ALCLOFENAC[+] AND ITS METABOLITES IN PLASMA OF HUMAN BEINGS

M.-J. SIMON, R. RONCUCCI, R. VANDRIESSCHE, K. DEBAST, P. DELWAIDE and G. LAMBELIN

Continental Pharma. Research Laboratories, 30 Steenweg op Haacht, 1830 Machelen (Belgium)

SUMMARY

A mass-fragmentography method is described for the determination of alclofenac (4-allyloxy-3-chlorophenylacetic acid) and two of its biodegradation products, namely 4-hydroxy-3-chlorophenylacetic acid and 4-(2,3-dihydroxy)-propyloxy-3-chlorophenylacetic acid in the same plasma sample at drug and metabolite concentrations actually found after therapeutic oral doses.

INTRODUCTION

Alclofenac is an antirheumatic drug used in clinical practice since 1969[1]. Metabolic studies conducted in man, rhesus monkey, dog, pig and rodents after oral intake revealed in the urine the presence of the unchanged drug (A) and three of its metabolites, *i.e.* its ester glucuronide (AG), its O-dealkylated (4-HCPA) and its allyldihydroxylated (DHA) derivatives. In the pig, the glycine conjugate of A (HAA) was also detected (Table I).

Conspicuous interspecific differences were noted in the relative amounts of the metabolites excreted: AG and free A were found to be the major urinary end-products in man and rhesus monkey whereas DHA and 4-HCPA were the main metabolites in pig and rodents (Table II).

The detection of these metabolites by mass spectrometry and the elucidation of their fragmentation patterns were greatly facilitated by the presence of a chlorine atom which gives rise to characteristic doublets in their spectra[2,3].

[+]Mervan[R], Continental Pharma s.a., Brussels, Belgium.

Table I

Metabolites of alclofenac

Structure		Name / formula
$CH_2=CH-CH_2-O-C_6H_3(Cl)-CH_2-COOH$	A	4-allyloxy-3-chlorophenylacetic acid $C_{11}H_{11}O_3Cl_1$ MW = 226.039
$CH_2=CH-CH_2-O-C_6H_3(Cl)-CH_2-CO-O-$(glucuronic acid: COOH, HO, OH, OH)	A-G	1'-β-D-glucuronic acid-4-allyloxy-3-chlorophenylacetate $C_{17}H_{19}O_9Cl_1$ MW = 402.071
$CH_2(OH)-CH(OH)-CH_2-O-C_6H_3(Cl)-CH_2-COOH$	DHA	4-(2,3-dihydroxy)propyloxy-3-chlorophenylacetic acid $C_{11}H_{13}O_5Cl_1$ MW = 260.040
$HO-C_6H_3(Cl)-CH_2-COOH$	4-HCPA	4-hydroxy-3-chlorophenylacetic acid $C_8H_7O_3Cl_1$ MW = 186.008
$CH_2=CH-CH_2-O-C_6H_3(Cl)-CH_2-CO-NH-CH_2-COOH$	HAA	(N-carboxymethyl)-4-allyloxy-3-chlorophenylacetamide $C_{13}H_{14}N_1O_4Cl_1$ MW = 283.061

Table II

Urinary metabolites of alclofenac in man and in different animal species

Species	Approx. % of the total urinary end-products		
	A + AG	DHA	4-HCPA
Man	72	11	8
Monkey (rhesus)	74	5	5
Pig	17	76	7
Rabbit	16	53	31
Rat	traces	50	30
Mouse	4	55	27

After therapeutic doses in human beings, urinary contents of A, 4-HCPA and DHA can be easily assayed by gas chromatography by the method described previously[4]. Although usually accounting for the highest amount of A excreted in the urine (AG/A = 2.3 after oral intake of the drug[3]), AG is not measured as such in biological fluids; the assessment of this metabolite is performed by measuring A freed by acid hydrolysis of AG.

On the other hand, determinations of plasma levels of A, 4-HCPA and DHA in human beings after therapeutic doses (usually 3 x 500 mg or 3 x 1 g per day) showed that concentrations of A are usually high enough to be assayed with accuracy by the above-mentioned gas chromatographic method whereas 4-HCPA and DHA cannot even be detected by the same technique. However, since after oral doses of A, these two compounds are always present in urine in concentrations each accounting roughly for 1/10 of that of the parent drug and its glucuronide, it can be predicted that trace amounts will be found in plasma. Therefore, to get further information on the pharmacokinetics of A in man, a more sensitive analytical procedure was devised for the determination of the three compounds in the same plasma sample.

Since these compounds showed appropriate fragments by electron impact-mass spectrometry and since internal standards were available, mass fragmentography appeared the most suitable method to be developed.

EXPERIMENTAL

Biological samples and reagents

Blood samples were obtained from a fully informed volunteer who received orally one single dose of ^{14}C-labelled alclofenac (one capsule 500 mg, 30.7 µCi). Plasma specimens were kept frozen until analysis. A TRI-CARB Packard (Model 3375) liquid scintillation counter was used for ^{14}C determinations. Only analytical grade reagents were used.

Table III shows the structures of the compounds used as internal standards for the mass fragmentography assay of A, 4-HCPA and DHA. Working solutions (in methanol) containing 10 µg of A, 1 µg of 4-HCPA and 0.1 µg of DHA per µl were prepared.

Extraction and derivatization procedures

To 4 ml of plasma, internal standards of A (100 µg), 4-HCPA (6 µg) and DHA (2 µg) as well as 6 ml 1 *M* HCl were added. Hydrolysis was performed by refluxing the mixture for 20 min; the cooled mixture was diluted with

Table III

Derivatives used for mass fragmentography of A, 4-HCPA, DHA and the corresponding internal standards

	METABOLITE		INTERNAL STANDARD	
	UNDERIVATIZED	DERIVATIZED	UNDERIVATIZED	DERIVATIZED
A	CH_2=CH-CH_2-O-(Cl-phenyl)-CH_2-COOH MW 226	CH_2=CH-CH_2-O-(Cl-phenyl)-CH_2-COOTMS MW 298	(CH_2=CH-CH_2)(CH_2=CH-CH_2-O)-(Cl-phenyl)-CH_2-COOH MW 266	(CH_2=CH-CH_2)(CH_2=CH-CH_2-O)-(Cl-phenyl)-CH_2-COOTMS MW 338
4-HCPA	HO-(Cl-phenyl)-CH_2-COOH MW 186	TMSO-(Cl-phenyl)-CH_2-COOTMS MW 330	HO-(Cl-phenyl)-COOH MW 172	TMSO-(Cl-phenyl)-COOTMS MW 316
DHA	CH_2(OH)-CH(OH)-CH_2-O-(Cl-phenyl)-CH_2-COOH MW 260	CH_2(OTMS) - CH(OTMS)-CH_2-O-(Cl-phenyl)-CH_2-COOME MW 418	CH_2(OH)-CH(OH)-CH_2O-(Cl-phenyl)-COOH MW 246	CH_2(OTMS) - CH(OTMS)-CH_2-O-(Cl-phenyl)-COOME MW 404

5 ml of distilled water and extracted twice with ethyl acetate (1:5, v/v). The extracts were then dried on anhydrous Na_2SO_4 and distilled in vacuum (5 mm Hg) at 30°. The residue was dissolved in 50 µl of 1,4-dioxan and divided into two 25-µl aliquots. The first was derivatized with 25 µl of bis-trimethylsilyltrifluoroacetamide (BSTFA; Pierce, Rockford, Ill., U.S.A.) containing 3% trimethylchlorosilane (TMCS; Pierce). Specimens of 1 µl were injected for the mass fragmentography determination of both A and 4-HCPA. The second 25-µl aliquot was mehylated with diazomethane in methanol. The dry residue was then dissolved in 25 µl of 1,4-dioxan and silylated with 25 µl of BSTFA containing 3% TMCS. For the mass fragmentography determination of DHA, 1 µl of this solution was injected onto the column.

Mass fragmentography

A Varian CH7 GC-MS system equipped with an automated multiple-ion selection (MIS) accessory was used. The mass spectrometer was interfaced to a Varian 620 L/100 computer for on-line collection of data. Table IV shows the operating conditions for the assay of A and its metabolites.

The derivatives used (Table V) were the mono-trimethylsilyl (mono-TMS) for A and the bis-trimethylsilyl (bis-TMS) for 4-HCPA; the bis--trimethylsilyl methyl ester derivative (bis-TMS-Me) of DHA was preferred to the corresponding tris-TMS derivative for which interference was encountered in the fragmentograms. Internal standards were derivatized with the same techniques used for the corresponding compound (Table III). The mass spectra of the derivatized metabolites as well as those of their corresponding internal standards are shown in Figs. 1, 2 and 3.

Suitable calibration curves were obtained in plasma extracts, by focusing the ions shown in Table V. Other ions, although of higher intensity, were discarded because of the interference encountered in plasma fragmentograms. This was done, for instance, for the ion at *m/e* 213 of the A-mono-TMS derivative.

Calibration curves were performed by hydrolizing 4-ml aliquots of blank plasma containing the three internal standards at the concentrations given above (extraction procedure) and increasing amounts of each of the three compounds to be assayed. Linear responses were obtained for various

Table IV

GC/MS operating conditions for mass fragmentographic assay of A, 4-HCPA and DHA

GC/MS VARIAN MAT CH7	A	4-HCPA	DHA
Column	Glass, 2 m x 2 mm	Glass, 2 m x 2 mm	Glass, 2 m x 2 mm
Liquid phase	1% OV-17	3% SE-30	3% SE-30
Support	80-100 mesh Supelcoport	80-100 mesh Supelcoport	80-100 mesh Supelcoport
Temperature:			
Injector port	275°	275°	275°
Column	iso 180^{0}	iso 190°	iso 230°
Separator	290°	290°	290°
EI source	290°	290°	290°
Ionization voltage	70 eV	70 eV	70 eV
Trap current	300 µA	300 µA	300 µA

Table V

Ions selected for the mass fragmentographic assay of A, 4-HCPA and DHA

Compound		Selected fragments	
		m/e	Proposed formula
A	(TMS)	298–300	M^+
		283–285	M^+-15
Int. stand.	(TMS)	323–325	M^+-15
4-HCPA	(TMS)	330–332	M^+
Int. stand.	(TMS)	301–303	M^+-15
DHA	(TMS-Me)	272–274	TMSO-⟨O⟩-CH_2-$COOCH_3$ (Cl)
Int. stand.	(TMS-Me)	258–260	TMSO-⟨O⟩-$COOCH_3$ (Cl)

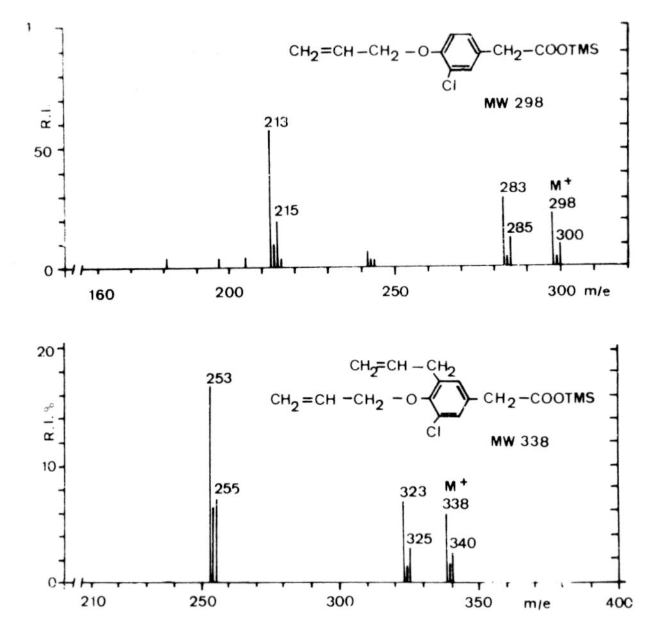

Fig. 1. Mass spectra of TMS derivatives of alclofenac (top) and the corresponding internal standard used for alclofenac mass fragmentographic assay (bottom). For conditions, see Table IV.

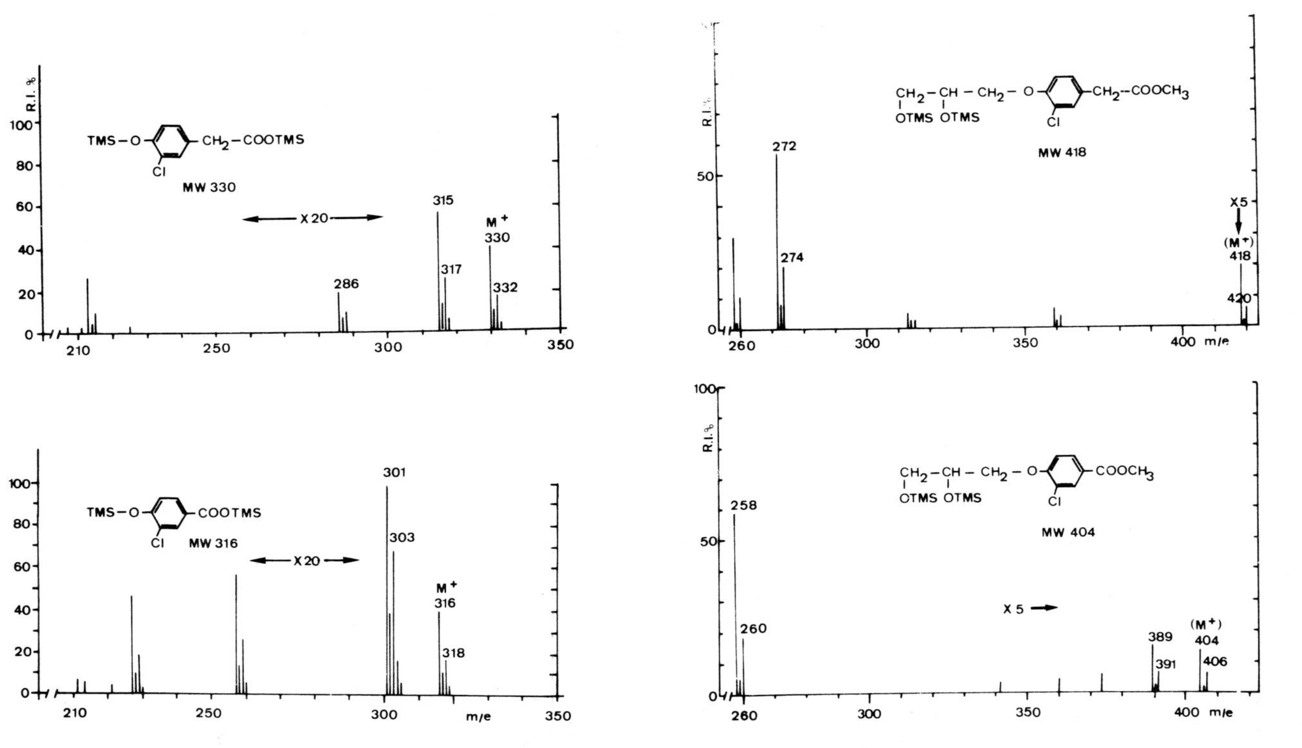

Fig. 2. Mass spectra of TMS derivatives of 4-HCPA (top) and the corresponding internal standard used for 4-HCPA mass fragmentographic assay (bottom). For conditions, see Table IV.

Fig. 3. Mass spectra of TMS derivatives of DHA (top) and the corresponding internal standard used for DHA mass fragmentographic assay (bottom). For conditions, see Table IV.

concentrations from 5 to 30 µg/ml for A, from 250 to 2500 ng/ml for 4-HCPA and from 125 to 500 ng/ml of plasma for DHA. Fig. 4 shows the calibration curves for 4-HCPA and DHA obtained by plotting the peak height ratio of the focused ions *versus* the concentrations in plasma of each of the compounds.

An example of a computerized mass fragmentogram obtained with 80 ng of 4-HCPA bis-TMS derivative and 60 ng of the corresponding internal standard bis-TMS derivative contained in 1 µl of plasma extract is shown in Fig. 5. Ions with *m/e* 330 and 332 (4-HCPA), and 301 and 303 (internal standard), were focused.

RESULTS AND DISCUSSION

The analytical procedure described was applied to plasma samples obtained from one female volunteer (JJ, 31 y, 68 kg) to measure A, 4-HCPA and DHA levels at different times after one single administration of 500 mg of ^{14}C-carbonyl-labelled alclofenac. The results are reported in Table VI and Fig. 6.

From these data it appears that 4-HCPA and DHA plasma levels are of the same order of magnitude and that, compared with those of the parent drug, they each account for about 1% (*versus* 10% in the urine). These low concentrations allow us to understand why these two metabolites were not detected by the gas chormatographic method described previously[4] which

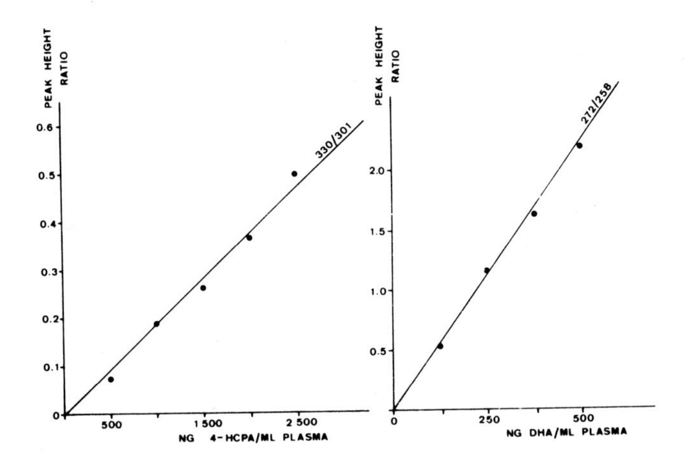

Fig. 4. Calibration curves for mass fragmentography determination of 4-HCPA (left) and DHA (right) in plasma. For conditions, see Table IV.

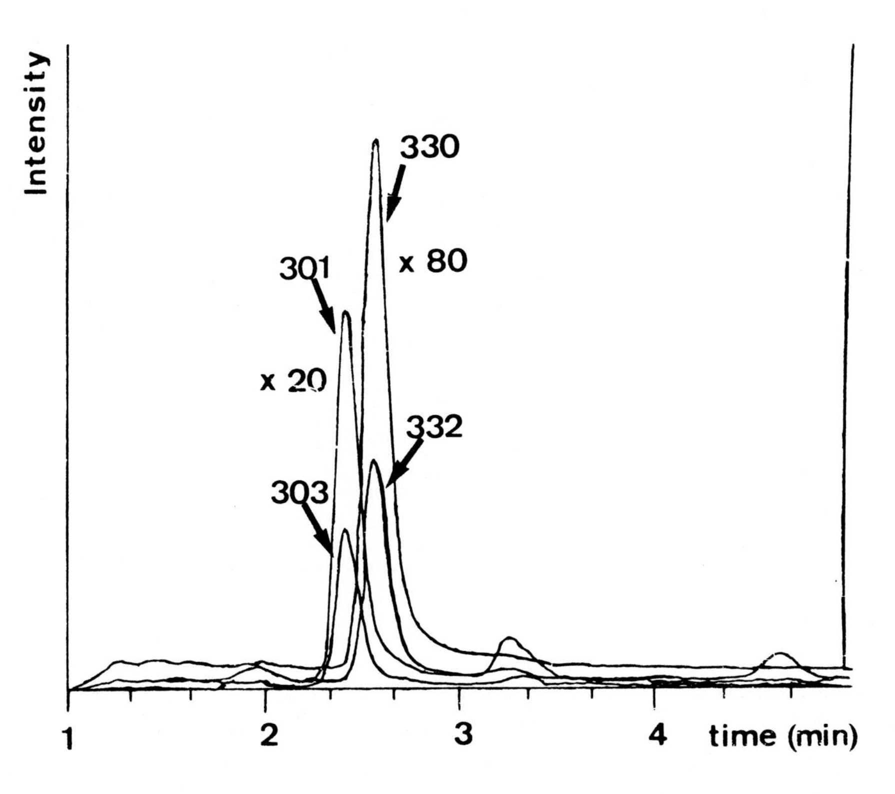

Fig. 5. Typical computerized mass fragmentogram obtained by analyzing a 1-μl aliquot of a plasma extract containing 80 ng of 4-HCPA (TMS derivative) and 60 ng of its internal standard (TMS derivative). Ions at *m/e* 330 and 332 for 4-HCPA, and 301 and 303 for its internal standard, were focused. Amplification factors were 80 and 20 for 4-HCPA and for its internal standard, respectively.

requires injection onto the column of at least 200 ng of each compound. The mass fragmentography assay we set up is at least 20 times more sensitive.

Table VI also shows that the sum of the metabolites accounts for nearly the whole radioactivity present in plasma.

From these preliminary results, it appears (Fig. 6) that the plasma peaks of 4-HCPA and DHA are both somewhat delayed compared to those of the parent drug and, with neglect of pharmacokinetic considerations, it also appears that the half-life of the metabolites is longer than that of the unchanged drug.

Table VI

^{14}C, A, 4-HCPA and DHA plasma levels in one volunteer after intake of 500 mg ^{14}C-alclofenac as a single oral dose (capsule)

Sampling time (h and min)	$^{14}C^{+}$ (μg A/ml)	A (μg/ml)	4-HCPA (ng/ml)	DHA (ng/ml)
0.05	0.02	0	0	0
0.10	6.65	2.65	87	0
0.15	17.20	9.81	230	0
0.20	22.60	15.40	251	0
0.30	34.94	29.29	486	215
0.45	40.12	34.13	523	246
1.00	47.36	40.50	530	297
1.30	48.38	46.43	560	431
2.00	48.55	42.86	600	471
4.30	25.30	20.50	500	270
6.00	13.21	8.57	205	178
12.00	1.79	0.94	70	0
24.00	0.34	0	31	0

$^{+}$Results obtained by liquid scintillation counting; 1 μg A = 136.4 dpm

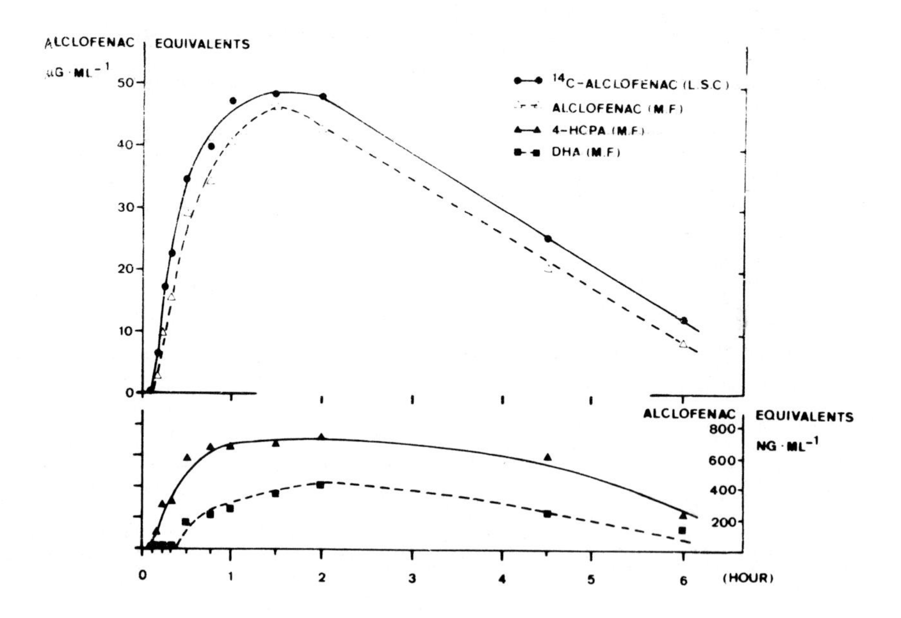

Fig. 6. Plasma levels of alclofenac and its metabolites in a female volunteer after oral intake of a tablet of 500 mg.

CONCLUSIONS

The analytical procedure described here allows accurate concurrent determinations of A and of its metabolites at low concentrations, the limit of sensitivity being 200 and 100 ng/ml of plasma for 4-HCPA and DHA, respectively. The experimental design reported here was set up in the belief that A plasma levels ranging between 1 and 50 µg/ml can be expected in man after one single oral intake of the drug; it now appears that far lower concentrations of A can be determined with accuracy by mass fragmentography. The analytical error of the method is estimated at less than 3% for A determinations and at about 5% for 4-HCPA and DHA in the range of concentrations encountered.

This mass fragmentography method will prove helpful for the quantitative determination of A and its metabolites in other tissues than plasma; these assays will probably contribute to improve the knowledge of the pharmacokinetics of the drug[5,6].

ACKNOWLEDGEMENT

This work was supported by a grant of the "Institut pour l'Encouragement de la Recherche Scientifique dans l'Industrie et l'Agriculture" of the Belgian Government (IRSIA, Convention 2459).

REFERENCES

1 N.P. Buu-Hoï, G. Lambelin, C. Gillet, J. Roba and M. Staquet, *Naturwissenschaften*, 56 (1969) 330.
2 R. Roncucci, M.-J. Simon, C. Gillet, G. Lambelin and M. Kaisin, in A. Frigerio (Editor), *Proceedings of the International Symposium on Gas Chromatography-Mass Spectrometry, Isle of Elba, 1972*, Tamburini Editore, Milano, 1972, p. 432.
3 R. Roncucci, M.-J. Simon, G. Lambelin and K. Debast, in A. Frigerio and N. Castagnoli (Editors), *Mass Spectrometry in Biochemistry and Medicine*, Raven Press, New York, 1974, p. 29.
4 R. Roncucci, M.-J. Simon, G. Lambelin, *J. Chromatogr.*, 62 (1971) 135.
5 R. Roncucci, M.-J. Simon, G. Lambelin, M. Staquet, C. Gillet, H. Van Cauwenberge, P. Lefebvre, J.C. Daubresse and N.P. Buu-Hoï, *Eur. J. Clin. Pharmacol.*, 5 (1971) 176.
6 M. Strolin-Benedetti, P. Strolin, R. Roncucci, M.-J. Simon, G. Lambelin and C. Nagant de Deuxchaisnes, *Eur. J. Clin. Pharmacol.*, 6 (1973) 261.

MEASUREMENT OF IMIPRAMINE AND DESIPRAMINE IN PLASMA, CEREBROSPINAL FLUID AND SALIVA OF PATIENTS WITH PRIMARY AFFECTIVE DISORDERS[+]

M. CLAEYS[++], G. MUSCETTOLA, F.K. GOODWIN and S.P. MARKEY

Laboratory of Clinical Sciences, National Institute of Mental Health, Bethesda, Md. 20014 (U.S.A.)

SUMMARY

A gas chromatographic-mass spectrometric (GC-MS) method for quantitative and simultaneous measurement of imipramine and desipramine is reported. The assay is based on stable-isotope dilution techniques, with deuterated analogues as internal standards. The determination of (compound)/(internal standard) ratios, which is central in this assay, is derived from selected ion recording data of $[MH]^+$ ions, formed upon chemical ionization with methane as reagent gas. Statistical data on the precision of the GC-MS assay are presented. Relevant clinical observations, obtained from measurement in samples from patients, include the findings of a significant correlation between concentrations in plasma and cerebrospinal fluid and a correlation between the level of imipramine and the extent of decrease in the probenecid-induced accumulation of 5-hydroxyindoleacetic acid, the major metabolite of serotonin.

INTRODUCTION

Imipramine (10,11-dihydro-5-(3',3'-dimethylaminopropyl)-5H-dibenz[*b,f*]-azepine) occupies an important place in the treatment of primary

---

[+] Abbreviations: TFA, trifluoroacetyl; TFAI, trifluoroacetylimidazole; CSF, cerebrospinal fluid; 5-HIAA, 5-hydroxyindoleacetic acid.

[++] Research Associate from the "Nationaal Fonds voor Wetenschappelijk Onderzoek", Belgium. Pressent address: Department of Organic Chemistry, State University of Ghent, Krijgslaan, 271, S.4, B-9000 Ghent, Belgium.

affective disorders. Earlier studies showed that the major metabolite in brain is desipramine (10,11-dihydro-5-(3'-methylaminopropyl)-5H-dibenz[*b,f*]-azepine), formed by N-demethylation in the liver[1]. Since this compound is also pharmacologically active, it appeared of importance to measure imipramine and desipramine, in order to correlate biological fluid levels with pharmacological effects. Both compounds exert their effects in the central nervous system through an inhibition of the re-uptake of biogenic amines at synaptic sites: imipramine is most effective on the serotonin system, whereas desipramine appears to be a more potent blocker of the re-uptake of noradrenaline[2].

This report gives a brief description of the gas chromatographic - - mass spectrometric (GC-MS) method, developed in our laboratory to allow simultaneous measurement of imipramine and desipramine[3]. Having applied this assay over an extended period of time, we were able to obtain information on the precision of the method. The results of this evaluation are also reported. The methodology was applied to the analysis of plasma, cerebrospinal fluid and saliva of patients treated with therapeutic doses of imipramine. The clinical findings presented emphasize the relevance of measurement of these drugs in the different biological fluids or tissues.

EXPERIMENTAL

Materials

Imipramine was supplied by Geigy (Ardsley, N.Y., U.S.A.) and desipramine by USV Pharm. (Tuckahoe, N.Y., U.S.A.). The 2,4,6,8-tetradeutero analogues of imipramine and desipramine were prepared by acid-catalysed exchange in $^{2}H_{2}O$; their preparation is described in detail in ref. 3. The 3'-trideuteromethyl analogue of imipramine was synthesized by N-trideuteromethylation, starting from desipramine and following the procedures described by McMahon *et al.*[4]. The 3'-trideuteromethyl analogue of desipramine was a gift from J. Shaw (National Heart and Lung Institute, Bethesda, Md. 20014, U.S.A.). The isotope distributions of the deuterated analogues, calculated by the use of a computer program (Labdet) developed by C.F. Hammer and A.J. Vlietstra at Georgetown University, Washington D.C., U.S.A., were as follows.

For $^{2}H_{4}$-imipramine: $^{2}H_{4}$, 59.5%; $^{2}H_{3}$, 3.3%; $^{2}H_{2}$, 6.2%; $^{2}H_{1}$, 0.8%; $^{2}H_{0}$, 0.2%.
For $^{2}H_{4}$-desipramine: $^{2}H_{4}$, 55.6%; $^{2}H_{3}$, 35.6%; $^{2}H_{2}$, 7.8%; $^{2}H_{1}$, 1.8%, $^{2}H_{0}$, 0.1%.

For $^{2}H_3$-imipramine: $^{2}H_3$, 88.4%; $^{2}H_2$, 9.8%; $^{2}H_1$, 1.8%.
For $^{2}H_3$-desipramine: $^{2}H_3$, 96.2%; $^{2}H_2$, 3.8%.
The derivatizing reagent was from Pierce (Rockford, Ill., U.S.A.).

Standards

Standard solutions of imipramine and desipramine were made by dissolving the hydrochloride salts of the amines in 0.1 *M* HCl to give about 0.2 mg/ml calculated as free base. From these solutions fresh 1/625 dilutions were made on each experimental day for assays in plasma and saliva, whereas 1/5000 dilutions were made for assays im CSF. All standards were made in drug-free biological fluid having concentrations ranging from 20 to 200 ng/ml for assays in plasma and saliva, and from 4 to 40 ng/ml for assays in CSF, and were prepared by multiple pipetting from a calibrated 100 µl pipette.

Extraction and derivatization

After addition of the internal standards and 1 ml 0.1 *M* sodium hydroxide, 2 ml of biological fluid was extracted twice with 3 ml hexane by horizontal shaking for 5 min in 15-ml glass-stoppered tubes. The hexane layers were evaporated to dryness under a stream of nitrogen. The recovery of the extraction procedure was estimated by using $^{2}H_4$-labelled analogues instead of radioactive tracer and was calculated for one concentration and for the assays in plasma and CSF. For the determination in plasma, concentration of 68.5 ng/ml for imipramine and of 76.5 ng/ml for desipramine were used and the recoveries were estimated to be 62 and 73%, respectively. For the determinations in CSF, concentrations of 21.4 ng/ml for imipramine and 23.9 ng/ml for desipramine were used; the recoveries were 95 and 87%, respectively. The dried extracts were dissolved in 20 µl of hexane, 2 µl TFAI was added and the mixture was reacted for 30 min at room temperature before analysis. Usually, 1-2 µl of this solution was injected for GC-MS analysis.

Instrumental conditions

A c.i. Finnigan gas chromatograph-mass sepctrometer interfaced with a 6000 data system was used. GC was performed on a 6 ft.x2 mm I.D. glass column, containing 1% OV-17 on Supelcoport 100-120 mesh with a helium flow-rate of 30 ml/min. Temperature were: injector, 250°; column, 230°; and GC-MS interface, 230°. The reagent gas, methane, was added as make-up gas. The MS conditions were: electron energy, 150 eV; emission current, 1 mA; continuous dynode electron multiplier voltage, 1.7 kV and pre-amp. range,

10^{-8} aV^{-1}. All selected ion records were obtained by recording $[MH]^+$ ions. Selected ion recording and subsequent determinations of peak areas were performed by using the data system and the revision H software.

Human samples

Depressed patients in hospital were studied in a metabolic research unit at the National Institute of Mental Health, specifically designed for the collection of behavioural and biochemical data on a longitudinal basis. The patients' symptoms were diagnosed by two psychiatrists and a social worker, according to Winokur's criteria for primary affective disorders[5]. The patients were drug-free for at least four weeks before the beginning of the study. Imipramine was administered in four divided doses ranging from 150 to 350 mg/day. Blood samples were drawn immediately before the administration of the morning dose, and plasma was obtained by centrifugation. Parotid saliva was collected after stimulation with a drop of lemon juice. Lumbar punctures were ordinarily performed at 09.00 h with the patients kept fasting and recumbent in bed during the preceding 9-12 h. Usually 10-12 ml CSF were collected. Biological samples were stored frozen until analyzed.

RESULTS AND DISCUSSION

Careful consideration of experimental parameters such as GC behaviour and MS characteristics of the compounds to be assayed is necessary when selected ion recording and isotope dilution techniques are applied to practical quantitative determinations. Suitable GC behaviour for desipramine, a secondary amine, was obtained after its conversion to a TFA derivative with TFAI before GC. Chemical ionization (c.i.) of imipramine and desipramine-TFA yielded intense $[MH]^+$ ions and was chosen in preference to electron ionization. C.i. has the advantage that it does not require the label at a specific position when ^2H-labelled analogues are used as internal standards for stable-isotope dilution. From the reagent gases methane, isobutane and ammonia, methane was selected on the basis of maximal sensitivity afforded by $[MH]^+$ ions. The c.i. (methane) mass spectra of imipramine and desipramine-TFA are presented in Figs. 1 and 2. Two different sets of ^2H-labelled internal standards were used: the 2,4,6,8--tetradeutero analogues, obtained by acid-catalyzed exchange in ^2H_2O and the more stable and more enriched 3'-trideutero analogues that became available, in the course of this work. Quantitation was achieved with a

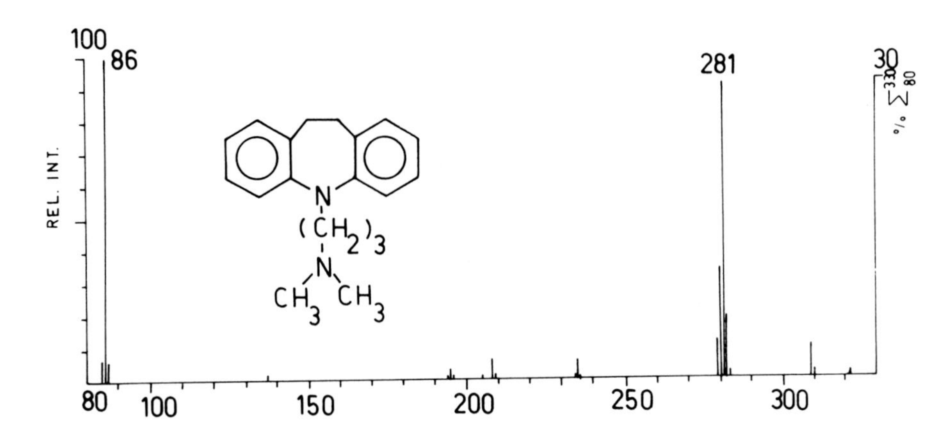

Fig. 1. C.i. (methane) mass spectrum of imipramine.

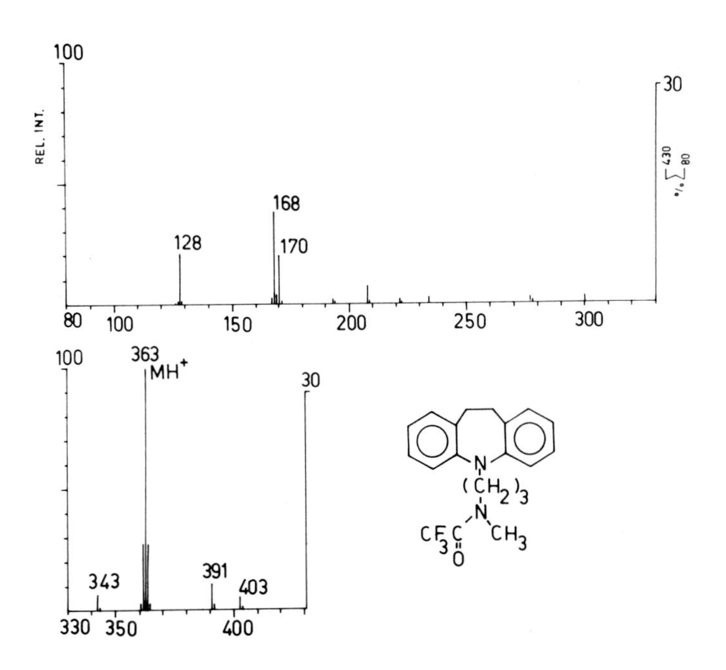

Fig. 2. C.i. (methane) mass spectrum of desipramine-TFA.

calibration curve, obtained by carrying known amounts of the compound together with a fixed amount of internal standard through the entire analysis and by plotting the (compound)/(internal standard) peak ratios, obtained from selected ion recording data, *versus* the concentration of compound. The calibration curve was obtained by unweighted linear regression analysis, and the concentrations of the unknowns were derived by using the regression line in reverse. A short computer program (in Basic)[+] was developed to facilitate these calculations. The selected ion records of a typical analysis and a calibration curve for imipramine are shown in Figs. 3 and 4. The precision of the results, derived for the unknowns, depends on the precision of the method, which in turn depends on the nature of the analyte and on the specification and performance of the mass spectrometer being used. It is agreed that the calibration curve should be

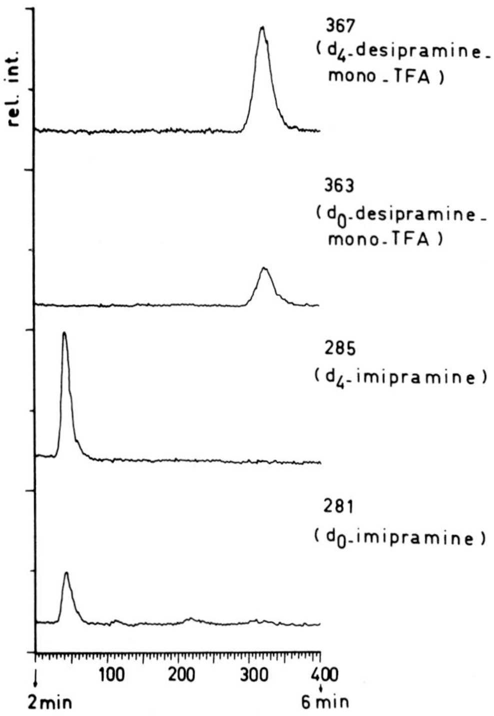

Fig. 3. Selected ion records of a derivatized plasma extract, to which $^2H_4$-labelled analogues were added.

[+] Available from the autors upon request.

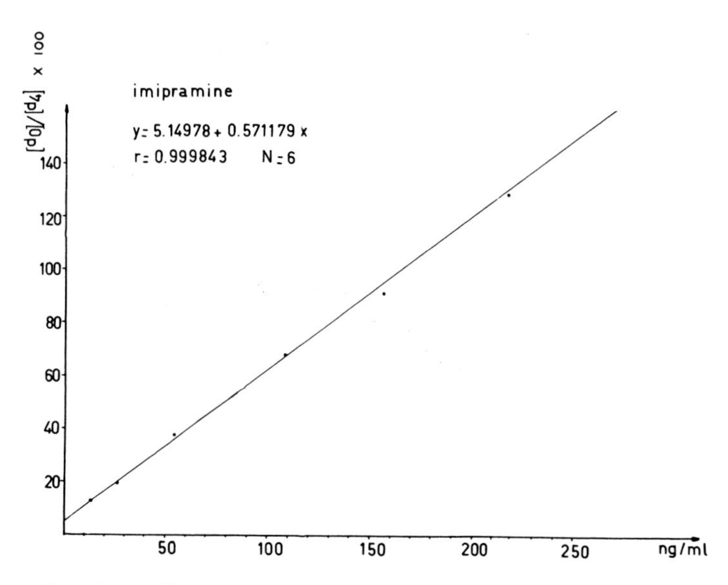

Fig. 4. Calibration curve with regression characteristics for imipramine in a typical experiment.

presented as a straight line[6], which can be obtained by linear regression analysis of the (compound)/(internal standard) peak ratio upon concentration of compound. The type of weighting, however, which should be applied in the performance of this analysis, is less obvious and solely depends on the error function. Errors are introduced into the analysis at two stages: firstly, in the sample manipulation step, and secondly, in the GC-MS analysis. Estimation of the relative contributions of both sources of error is of interest to provide an idea about their impact on the precision, whereas evaluation of their dependancy upon the concentration allows one to determine the type of weighting which has to be applied in the weighting of the data for linear regression analysis. In this work we report the preliminary results of this evaluation for the determination of imipramine and desipramine in plasma and using 2,4,6,8-tetradeutero analogues. The variances due to GC-MS analysis, $s^2_{\text{GC-MS}}$, and due to sample manipulations, s^2_{sample}, were derived by standard statistical techniques[7] and are listed in Tables I and II. A more detailed report will be published elsewhere[8]. From the results in Table I it seems that $s^2_{\text{GC-MS}}$ is fairly independent of the standard concentration; therefore, a better estimate for $s^2_{\text{GC-MS}}$ can be obtained by averaging the $s^2_{\text{GC-MS}}$ values. The average values for $s^2_{\text{GC-MS}}$ divided by the average slopes of the regression line result in the errors expressed in ng/ml and are 2.4 ng/ml for the imipramine and 1.7 ng/ml for

Table I

Variances due to the GC-MS analytical system, s^2_{GC-MS}, when $^2H_4$-labelled analogues were used as internal standards for imipramine and desipramine

Compound	Concentration of std (ng/ml)	s^2_{GC-MS}	Degrees of freedom	Average slope of regression line ($\bar{b}$, ng/ml)	$\bar{s}/\bar{b}$ (ng/ml)
Imipramine	17.1	0.20	26	0.254	2.4
	34.2	0.23	36		
	51.3	0.24	18		
	68.5	0.35	18		
	102.7	0.41	37		
	137.0	0.82	19		
	171.0	0.45	15		
	188.1	0.28	20		
		s^2_{GC-MS} = 0.37			
Desipramine	18.7	0.22	30	0.51	1.7
	37.4	0.44	18		
	56.1	0.39	21		
	74.8	1.03	19		
	112.2	0.51	36		
	149.6	1.25	19		
	187.0	1.26	30		
	205.7	1.01	25		
		s^2_{GC-MS} = 0.76			

the desipramine assay. In Table II, some values for s^2_{sample} are negative, others would probably not be significant according to an F test. Anyway, s^2_{sample}, appears to be independent of the standard concentration. The values for $\bar{s}_{sample}/\bar{b}$ are 1.1 ng/ml for both imipramine and desipramine and are lower than those obtained for $\bar{s}_{GC-MS}/\bar{b}$. For both assays about two-thirds of the standard deviation about the regression line can be explained as being due to GC-MS and sample manipulations[8]. Both these contributions are independent of the standard concentration, implying that an unweighted linear regression analysis can be applied.

We were mainly interested in measuring anti-depressant levels in several body compartments, such as plasma, CSF and saliva, and we started our practical applications by measuring plasma levels in patients under imipramine therapy. Many questions about the clinical effects of the

Table II

Variances due to sample manipulations, s^2_{sample}, when $^2H_4$-labelled analogues were used as internal standards for imipramine and desipramine

Compound	Concentration of std (ng/ml)	Peak ratio x100 (average)	s^2_{sample}	Average slope of regression line ($\bar{b}$, ng/ml)	$\bar{s}/\bar{b}$ (ng/ml)
Imipramine	34.2	13.3	0.03	0.254	1.1
	68.5	22.7	0.04		
	102.7	31.3	0.33		
	137.0	39.0	-0.05		
	171.0	48.6	0.04		
		s^2_{sample}	= 0.078		
Desipramine	19.1	11.4	0.27	0.51	1.1
	38.2	21.7	0.37		
	76.5	41.2	-0.26		
	114.7	60.9	0.11		
	153.0	81.1	1.10		
	191.0	99.0	0.38		
		s^2_{sample}	= 0.34		

tricyclics remain unanswered, and controversy still exists about the relation between plasma levels and clinical response. Studies on nortriptyline indicated a response between 40 and 175 ng/ml[9]; other studies have failed to demonstrate a relationship[10]. In our results for the steady-state levels, summarized in Fig. 5, a large variability in the levels can be noticed: there is a 5-fold difference for imipramine, whereas the range for desipramine levels is 3-fold. For 2 out of 8 patients, the steady-state levels of imipramine were higher than those of desipramine, a finding also observed by Belvedere *et al.*[11].

The CSF levels were measured in the same group of patients; the mean values together with the observed ranges are given in Table III. The low CSF levels could be measured by our technique down to 3 ng/ml with a standard error of 0.3 ng/ml. The CSF mean value was 10.6 ng/ml for imipramine and 15.0 ng/ml for desipramine, whereas the respective values in plasma were 118 and 112 ng/ml. The ratio (imipramine)/(desipramine) was lower in CSF, a result that can be explained by the lower protein-binding capacity of desipramine[12]. The correlations between plasma and CSF levels, presented in Fig. 6, were highly significant and indicate that the

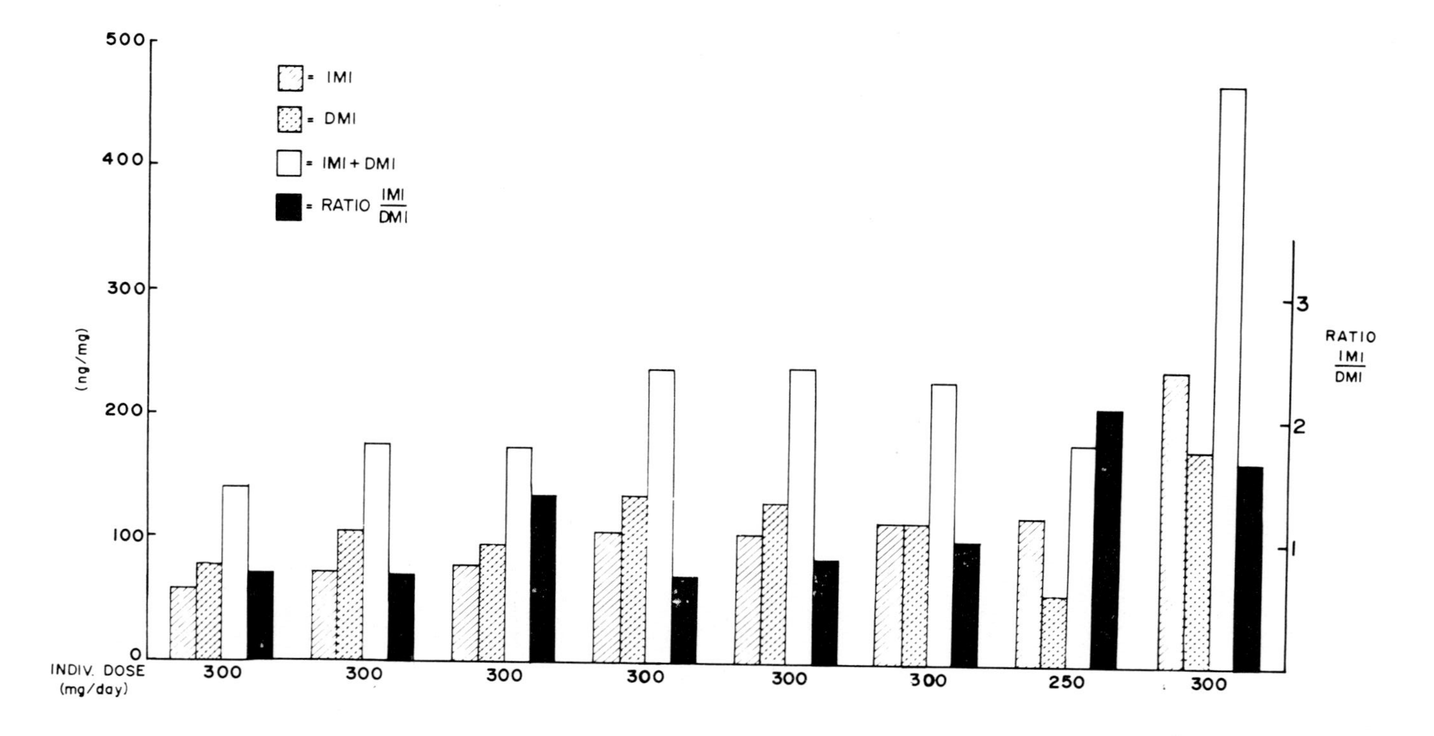

Fig. 5. Inter-individual differences in the steady-state plasma levels of imipramine and desipramine.

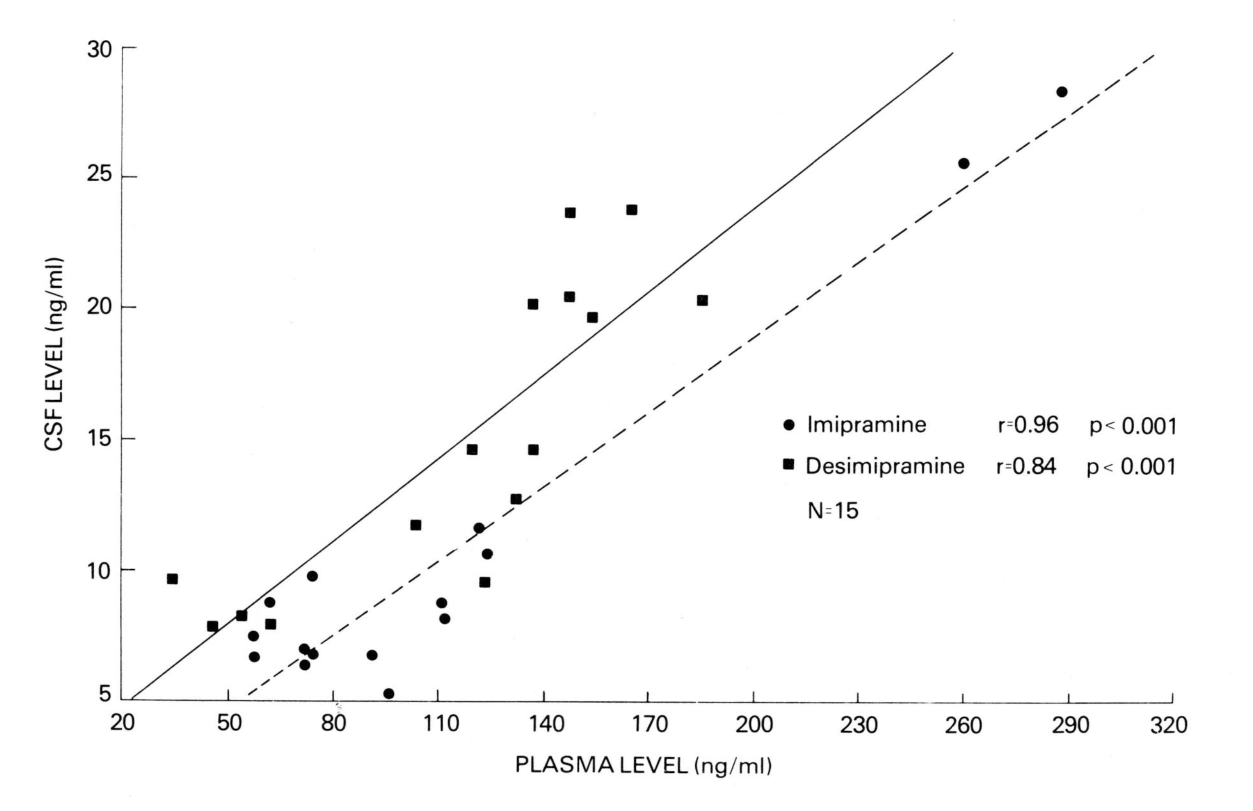

Fig. 6. Correlations between levels of plasma and CSF imipramine and desipramine.

Table III

Imipramine and desipramine levels in CSF

Data expressed as mean ± S.E.M.

	Imipramine	Desipramine
Mean	10.60 ± 1.77	4.95 ± 1.51
Range	5.8 - 28.4	7.8 - 23.7
% of plasma	9.36 ± 0.24	12.20 ± 0.79
Range	5.49 - 13.18	7.69 - 15.21

measurement of plasma levels, although it reflects total concentration of free and protein-bound drug, is a valid index for the active drug concentration in the central nervous system. The availability of plasma and CSF data on the same patients gave us the opportunity to evaluate the controversial problem of plasma protein-binding. We can assume that the CSF concentration is the free fraction of the drug, because only this fraction can cross the blood-brain barrier and because of the very low levels of protein present in the CSF. Expressing the imipramine and desipramine value in CSF as percentage of the respective values in plasma for each patient, we obtained a mean value for the percentage of free drug of 9.3% for imipramine and of 12.2% for desipramine, with a 2-fold range for both compounds (Table III). These results are in agreement with those obtained by Borga *et al.*[12], who found small inter-individual differences for the protein-binding of anti-depressant drugs. In contrast, an important difference in the protein-binding was found by Glassman *et al.*[13] and has been suggested as the cause of the large inter-individual differences in the response to antidepressant drugs.

Animal studies have shown that tertiary amines have more effect on the neuronal re-uptake of serotonin than they have on noradrenaline re--uptake[2]; on the basis of this, a correlation could be expected between imipramine levels and the degree of change in serotonin metabolism, which was measured by probenecid-induced accumulation of 5-HIAA[14]. Therefore, imipramine and desipramine levels were correlated with the decrease in accumulation of 5-HIAA; the results, given in Table IV, give some tentative confirmation that the effect of the drug on serotonin metabolism relates to imipramine rather than to desipramine.

Table IV

Correlation between imipramine and desipramine levels and the imipramine--induced change in probenecid accumulation of 5-HIAA in CSF (N = 8)

	Plasma		CSF	
Imipramine	r = 0.73	$p < 0.05$	r = 0.65	$p < 0.1$
Desipramine	r = -0.16	N.S.[+]	r = -0.46	N.S.[+]

[+] N.S., not significant.

Measurements in parotid saliva, obtained by acid stimulation, were carried out to evaluate whether this biological fluid could provide an easily accessible medium for monitoring drug concentrations in patients receiving imipramine. The salivary excretion of other drugs has been the subject of several investigations; a good correlation between serum and mixed saliva concentrations was demonstrated for theophylline[15], digoxin[16] and phenytoin[17]. In this work, imipramine and desipramine were measured in samples of plasma and saliva, collected simultaneously on various occasions. The correlation coefficients were calculated between the levels found in the two media and were 0.86 (N = 15; $p < 0.001$) and 0.85 (N = 15; $p < 0.001$) for imipramine and desipramine, respectively. The highly significant correlations suggest that levels found in parotid saliva, obtained upon acid stimulation, can be used as indices for plasma levels.

CONCLUSION

In conclusion, in this investigation, a sensitive and precise technique was developed for simultaneous measurement of imipramine and desipramine. The drugs were measured in samples of plasma, CSF and saliva from patients under imipramine therapy and significant correlations were found between the drug levels in plasma and CSF and between those in plasma and saliva. In our limited group of patients (N = 8) no correlation was found between levels and clinical response. This negative result could be explained by the fact that the majority of the patients happened to be non--responders, having low pretreatment levels of 5-HIAA in their CSF. The low 5-HIAA levels indicate a deficiency in the serotonergic system and suggest that these patients may have profited better from another therapy, such as amitriptyline, which leads to less demethylated product and is more effective on the neuronal re-uptake of serotonin. The correlation between the imipramine concentrations and the extent of decrease in the probenecid- - induced accumulation of 5-HIAA almost reached significance and suggests that the results should be integrated with biochemical data, such as levels of amine metabolites, in order to do meaningful clinical studies on the relationship between levels of anti-depressant drugs and pharmacological response.

ACKNOWLEDGEMENTS

We thank Dr. C.F. Hammer from Georgetown University, Washington, D.C., for making available his Labdet program, Mr. J.W. Morris from Finnigan Corporation for technical assistance, Dr. J. Shaw from the National Heart and Lung Institute for a gift of $^{2}H_{3}$-labelled desipramine and Dr. I.J. Kopin from this laboratory for his interest in this work.

REFERENCES

1 J. Christiansen and C.F. Gram, *J. Pharm. Pharmacol.*, 25 (1973) 604.
2 A. Carlsson, H. Corrodi, F. Fuxe and T. Hokfeld, *Eur. J. Pharmacol.*, 5 (1969) 357.
3 M. Claeys, G. Muscettola and S.P. Markey, *Biomed. Mass Spectrom.*, in press.
4 R.E. McMahon, F.J. Marshall, H.W. Culp and W.M. Miller, *Biochem. Pharmacol.*, 12 (1963) 1207.
5 G.W. Winokur, P.J. Clayton and T. Reich, *Manic-Depressive Illness*, C.V. Mossby Co., St.Louis, 1969.
6 J.F. Pickup, *Ann. Clin. Biochem.*, 13 (1976) 306.
7 O.L. Davies and P.L. Goldsmith, *Statistical Methods in Research and Production*, Oliver and Boyd, Edinburgh, 4th ed., 1972.
8 M. Claeys, S.P. Markey and W. Maenhaut, *Biomed. Mass Spectrom.*, to be submitted for publication.
9 P. Kragh-Sõrensen, M. Asberg and C. Eggert-Hanssen, *Lancet*, i (1973) 113.
10 G.D. Burrows, B. Davies and B.A. Scoggins, *Lancet*, ii (1972) 619.
11 G. Belvedere, L. Burti, A. Frigerio and C. Pantarotto, *J. Chromatogr.*, 111 (1975) 313.
12 O. Borga, D.L. Azarnoff, G.P. Forshell and F. Sjõqvist, *Biochem. Pharmacol.*, 18 (1969) 2135.
13 A.H. Glassman, M.J. Hurwic and J.M. Perel, *Amer. J. Psychiat.*, 130 (1973) 1367.
14 N.R. Tamarkin, F.K. Goodwin and J. Axelrod, *Life Sci.*, 9 (1970) 1397.
15 R. Koysooko, E.F. Ellis and G. Levy, *Clin. Pharmacol. Ther.*, 15 (1974) 454.
16 W.J. Jusko, L. Gerbracht, L.H. Golden and J.R. Koup, *Res. Commun. Chem. Pathol. Pharmacol.*, 10 (1975) 189.
17 F. Bochner, W.D. Hooper, J.M. Sutherland, M.J. Eadie and J.H. Tyrer, *Arch. Neurol.*, 31 (1974) 57.

MASS SPECTROMETRIC ANALYSIS OF PHENCYCLIDINE IN BODY FLUIDS OF INTOXICATED PATIENTS

DENIS C.K. LIN and RODGER L. FOLTZ

Battelle, Columbus Laboratories, Columbus, Ohio 43201 (U.S.A.)

and

ALLAN K. DONE, REGINE ARONOW, EDGARDO ARCINUE and JOSEPH N. MICELI

Children's Hospital of Michigan, Detroit, Mich. 48201 (U.S.A.)

SUMMARY

A comparison was made of the sensitivities achieved with several different modes of ionization for the quantitative analysis of phencyclidine by the technique of selected ion recording. Chemical ionization with a reagent-gas mixture of methane and ammonia gave the highest sensitivity primarily because of the total absence of fragmentation. Chemical ionization with methane alone as the reagent gas was only 20% as sensitive, whereas electron-impact ionization was about 60% as sensitive. The analysis of serial samples of body fluids from phencyclidine-intoxicated patients shows that administration of appropriate agents to bring about acidification of the urine is an effective therapeutic technique.

INTRODUCTION

1-(1-Phenylcyclohexyl)piperidine (I), otherwise known as phencyclidine or PCP, was originally used as an analgesic anaesthetic drug for humans but, owing to its side effects, it is now used exclusively as an animal tranquilizer. In North America, phencyclidine is widely available on the illicit market, often sold as "Angel Dust", "Peace Pill", or under the name of other hallucinogens such as mescaline, LSD or THC. Phencyclidine is now recognized as a potent and dangerous drug[1,2]. In some areas of the U.S.A. it is the major cause of drug intoxications[3]. Recently we reported[4] a method for quantification of phencyclidine in body fluids. The method uses the combination of gas chromatography and methane chemical ionization mass spectrometry with the technique of selected ion recording. Pentadeuterated phencyclidine is used as the internal standard. The sensitivity of the method is such that concentrations of phencyclidine as low as 1 ng/ml can be measured in 1-ml samples of body fluids. The same paper reported the identification of two human metabolites of phencyclidine, 4-phenyl-4-piperidinocyclohexanol (II) and 1-(1-phenylcyclohexyl)-4--hydroxylpiperidine (III). A third metabolite was tentatively identified as 1-(1-phenyl-4-hydroxycyclohexyl)-4-hydroxypiperidine (IV) in the urine of rhesus monkeys. Since then several reports have been published concerned with the analysis of phencyclidine and identification of its metabolites[5-7].

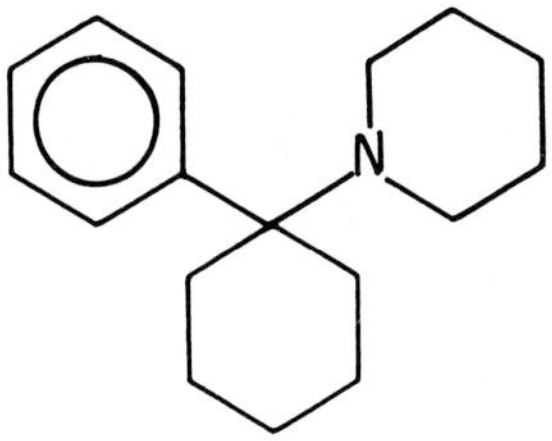

N

I

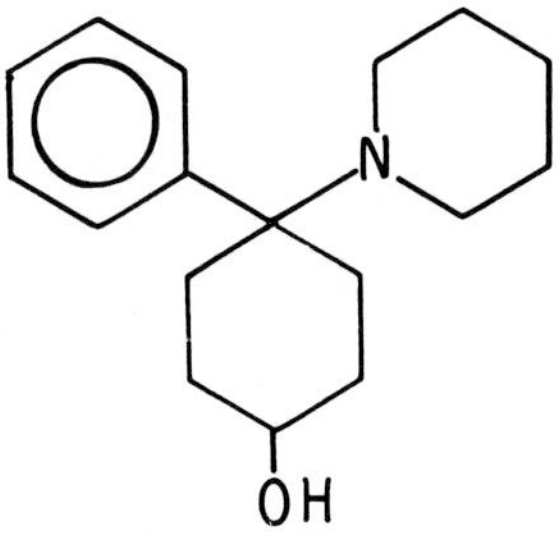

N
OH

II

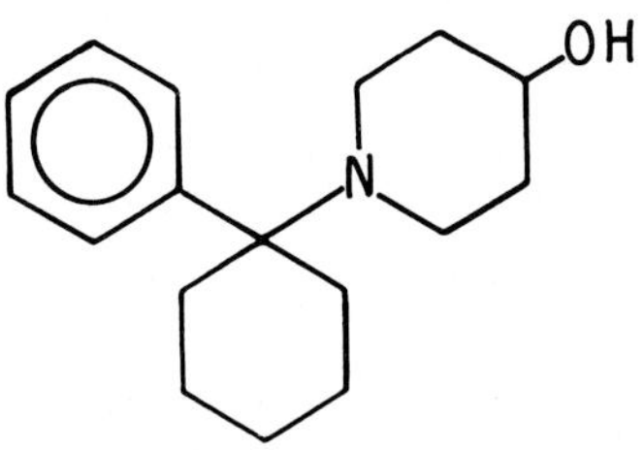

N
OH

III

N
OH
OH

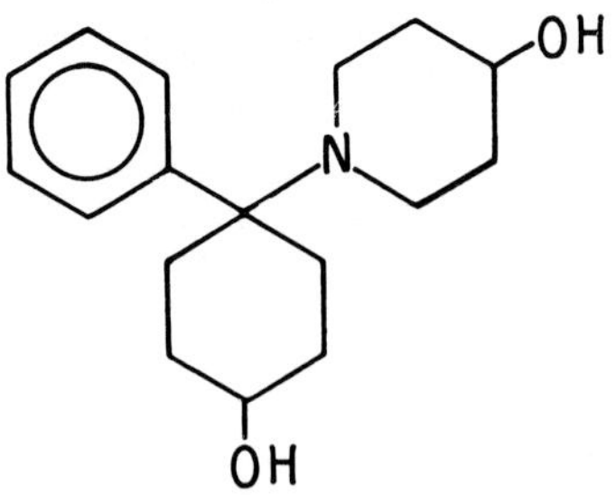

IV

This paper reports the results of our most recent work on phencyclidine. A comparison of several modes of ionization showed that better sensitivity and selectivity can be achieved in the quantitative analysis of phencyclidine by adding ammonia to the methane reagent gas. Also, phencyclidine concentrations have been measured in the body fluids of drug-intoxicated patients. These results indicate that acidification of the patients' urine causes rapid elimination of the drug and is a far more effective treatment for phencyclidine intoxication than is dialysis.

EXPERIMENTAL PROCEDURE

Comparison of sensitivities in electron impact (EI) and chemical ionization (CI)

Two quadrupole gas chromatography-mass spectrometry (GC-MS) systems were used in the study, one optimized for EI and the other for CI. We first determined the *m/e* value most suitable for selected ion recording in each of the modes of ionization. We then made repetitive 1-μl injections of either standard solutions of biological extracts containing known concentrations of phencyclidine, and measured the maximal ion current at the selected *m/e* value. For each series of runs the voltages of the ion source were adjusted to give the best performance for the particular mode of ionization being used. All other parameters were kept as constant as possible, including the GC conditions, the filament emission and the electron multiplier gain. The GC-MS interface in the EI system consisted of a Wolen diverter valve and a glass jet separator. The interface in the CI system consisted of a 1/8-in. stainless-steel tube and a specially constructed mechanical diverter valve. Because comparable relative sensitivities were obtained when the samples of drug were introduced via the direct probe inlet, we are confident that the GC-MS interfaces did not significantly affect the relative sensitivities.

In the CI system the GC carrier gas also served as the CI reagent gas. However, if necessary, we added a second reagent gas through a separate inlet into the ion source. Thus, to obtain ammonia-mediated chemical ionization, we bled NH_3 into the ion source to give a partial pressure (100 to 150 microns) sufficient to completely suppress the carrier gas ions.

Quantitative analysis of serial samples of body fluids

Our earlier report[4] described in detail the analytical procedure used to measure phencyclidine in samples of body fluids. Since the concentrations in the samples involved were relatively high, it was not necessary to increase the sensitivity by bleeding ammonia into the ion source as described in our comparison study.

RESULTS AND DISCUSSION

Comparison of sensitivities using EI and CI

The EI mass spectrum of phencyclidine (Fig. 1, top) shows an abundant ion at *m/e* 200 resulting from loss of C_3H_7 from the cyclohexyl ring. Since the deuteruim atoms in the internal standard are on the aromatic ring[4], the analogous ion has an *m/e* value of 205; therefore, the ion at *m/e* 200 is perfectly satisfactory for selected ion recording. In the methane CI mass spectrum (Fig. 1, middle), the $M^{+\cdot}$ was more intense than the MH^+, suggesting that charge transfer competes effectively with proton transfer when CH_4 is used as the reactant gas. In the methane and ammonia CI mass spectrum (Fig. 1, bottom), the protonated-molecule ions gave the most intense peak. Fragmentation was virtually absent. Relative responses for prominent ions in the various modes of ionization are compared in Table I. The values in the last column of Table I indicate that the best sensitivity was achieved with a mixture of methane and ammonia as the reagent gas. Surprisingly, EI gave about 3 times as much sensitivity as methane CI. However, if maximal sensitivity is required, EI, CI(CH_4) and CI(CH_4+NH_3) should be evaluated with respect to the extent of interference from background and endogenous materials. Our experience indicates that

Table I

Relative responses for prominent ions in the EI and CI mass spectra of phencyclidine

Ionization method	*m/e* monitored	% of Sample ion current	Relative response per unit weight of drug
EI	200 ($M^{+\cdot}-C_3H_7$)	14	59
CI (CH_4)	243 ($M^{+\cdot}$)	24	20
CI (CH_4+NH_3)	244 (MH^+)	73	100

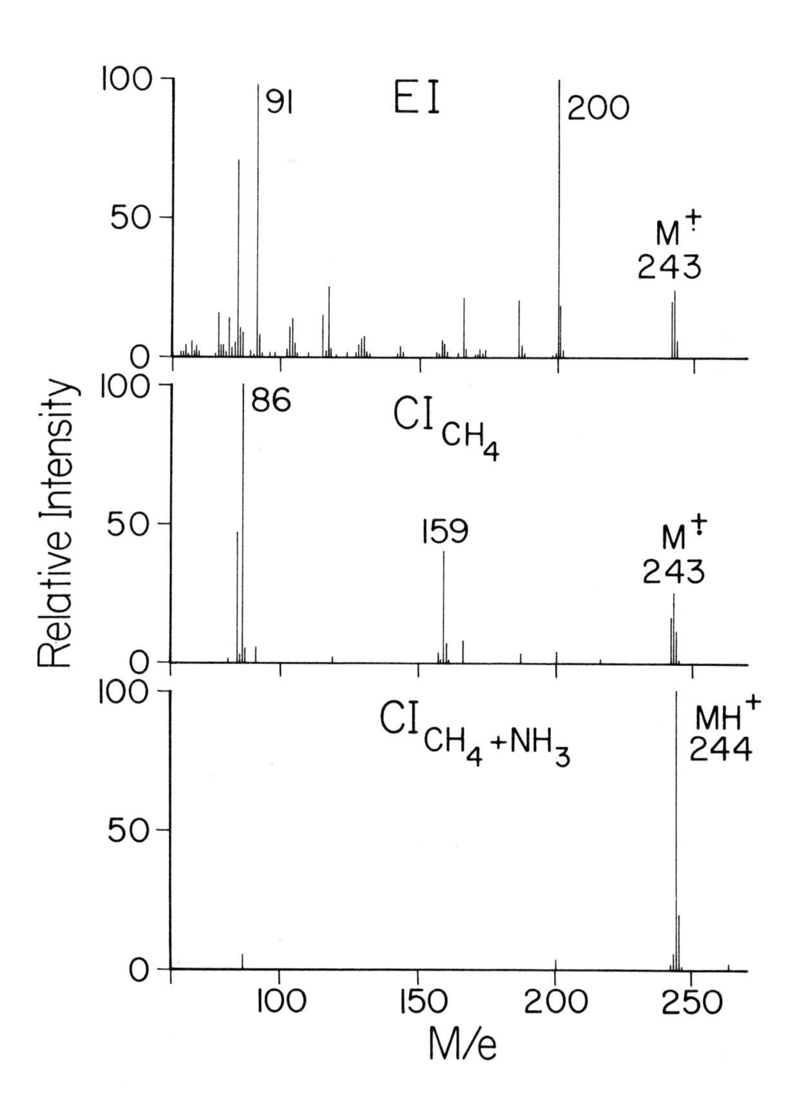

Fig. 1. Mass spectra of phencyclidine.

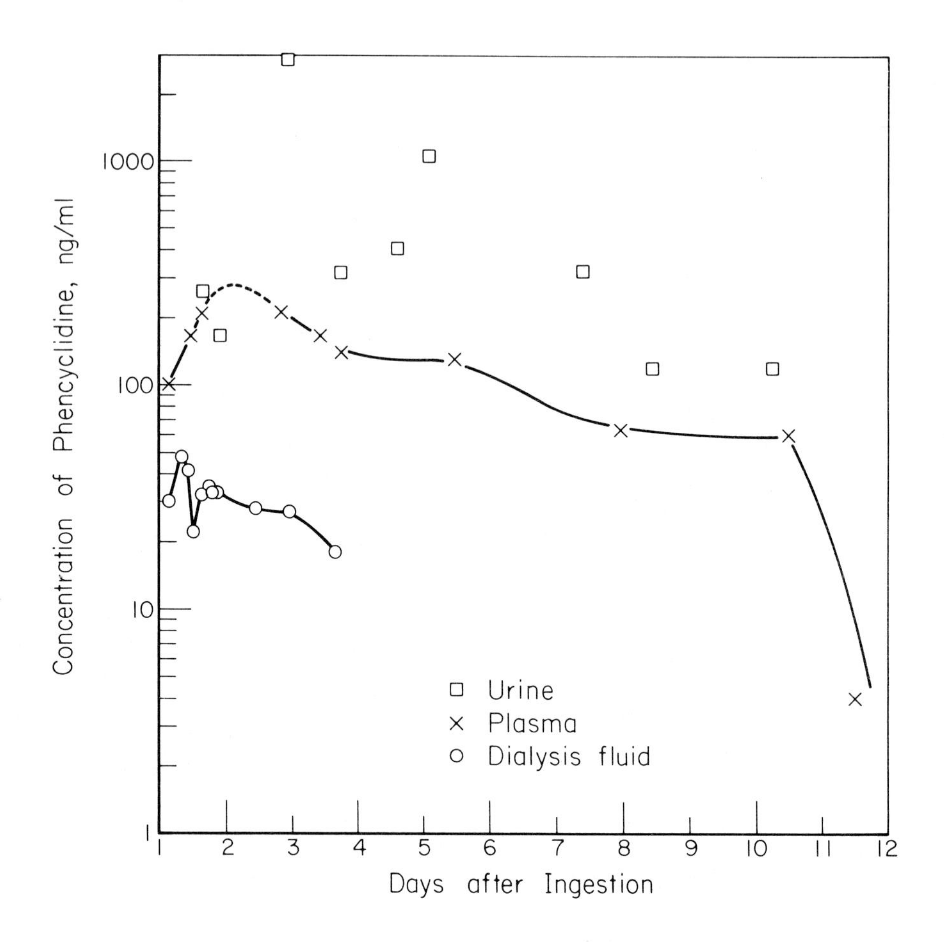

Fig. 2. Concentration of phencyclidine in samples of body fluids from patient No. 1.

CI(CH_4+NH_3) is more selective and, therefore, phencyclidine analyses using this reagent gas are less subject to interferences from other components of the extracts.

Quantitative analysis of serial samples of body fluids

In cases of intoxication, phencyclidine exerts a psychotropic effect which varies remarkably in duration from one patient to another. To examine this variance, we analyzed serial samples of body fluids from intoxicated patients. The body fluids included urine, gastric contents and plasma. To test whether phencyclidine is dialyzable, we also analyzed for the amount of phencyclidine eliminated in the dialysis fluids used in the treatment of some patients.

The plasma concentration of phencyclidine in patient No. 1 remained at a high level for more than 10 days (Fig. 2). Furthermore, peritoneal dialysis, begun on Day 1, did not stop the rise in plasma concentration. The low cencentration of the drug in the dialysis fluids along with the continued rise in plasma level indicates that dialysis is not an effective means of treating phencyclidine intoxications. The remarkably long retention of phencyclidine by this patient appears inconsistent with the plasma half-life of about 11 h reported by Marshman *et al.*[7]. However, it is known that the recovery time for phencyclidine intoxication varies from a few hours to days or even weeks. This suggests a wide variation in patients' ability to eliminate the drug. Table II relates the amount of phencyclidine eliminated in the urine of the same patient with the pH of the urine. Clearly the pH of the urine had a profound effect on the rate of elimination of the drug.

To study further the dependence of phencyclidine excretion on pH of the urine, samples of urine from a second patient were analyzed. The results (Table III) again show that the acidic urine samples contained larger amounts of the drug than did those with higher pH values.

The data obtained from patients Nos. 1 and 2 indicate that the acidification of urine should be an effective means to treat phencyclidine--intoxicated patients. Since this initial finding, we have successfully treated several severe cases of phencyclidine intoxication by administration of agents to produce acidification of the urine. We will report the results, along with a more detailed analysis of the data on our earlier patients, in a treatment-oriented medical journal.

Table II

Data on urine samples from patient No. 1

Date	Time of collection	Volume (ml)	pH	PCP concentration in urine (ng/ml)	Total PCP excreted (µg)
2/11	10A-4P	110	8.2	260	28.6
2/11	4P-10P	97	8.8	168	16.3
2/12	5P-11P	108	5.6	2900	313.2
2/13	12N-6P	102	8.2	320	32.6
2/14	9A-3P	105	8.4	407	42.7
2/14	9P-3A	94	6.4	1050	98.7
2/17	4A-10A	103	7.0	320	33.0
2/18	5A-11A	112	7.1	118	13.2
2/20	12M-6A	115	5.2	117	13.5

Table III

Data of urine samples from patient No. 2

Date	Time of collection	Volume (ml)	pH	PCP concentration in urine (ng)	Total excreted (µg)
2/16	12N-6P	51	7.42	181	9.2
2/17	6A-12N	52	8.46	50	2.6
2/17	12N-6P	104	7.40	144	15.0
2/18	6A-12N	92	6.20	394	36.3
2/19	12M-6A	53	5.07	574	30.4
2/19	12N-6P	58	6.28	131	7.6
2/19	12N-6P	84	6.90	60	5.0
2/21	12M-6A	41	5.34	119	4.9

ACKNOWLEDGEMENTS

This work was supported in part by the National Institute on Drug Abuse, Grant No. DA 00919. We would like to thank the nursing staff of the Children's Hospital of Michigan for their assistance in this study.

REFERENCES

1 J.W. Eastman and S.N. Cohen, *J. Amer. Med. Ass.*, 231 (1975) 1270.
2 B. Fauman, F. Baker and L.W. Coppleson, *J. Amer. Coll. Emerg. Physicians*, 4 (1975) 223.
3 J.P. Horwitz, E.B. Hills, D. Andrzejewski, W. Brukwinski, J. Penkala and S. Albert, *J. Amer. Med. Ass.*, 235 (1976) 1708.
4 D.C.K. Lin, A.F. Fentiman, Jr., R.L. Foltz, R.D. Forney, Jr. and I. Sunshine, *Biomed. Mass Spectrom.*, 2 (1975) 206.
5 C. Helisten and A.T. Shulgin, *J. Chromatogr.*, 117 (1976) 232.
6 K. Bailey, D.R. Gagne and R.K. Pike, *J. Ass. Offic. Anal. Chem.*, 59 (1976) 81.
7 J.A. Marshman, M.P. Ramsay and E.M. Sellers, *Toxicol. Appl. Pharmacol.*, 35 (1976) 29.

QUANTITATIVE DETERMINATION OF DOXEPINE AND DESMETHYLDOXEPINE IN RAT PLASMA BY GAS-LIQUID CHROMATOGRAPHY - MASS FRAGMENTOGRAPHY

A. FRIGERIO, C. PANTAROTTO, R. FRANCO, G. BELVEDERE, R. GOMENI and P.L. MORSELLI

Istituto di Ricerche Farmacologiche "Mario Negri", Via Eritrea 62, 20157 Milan (Italy)

The methods to determine the levels of tricyclic antidepressant drugs generally lack sensitivity and/or specificity, particularly when applied to biological specimens. The method reported here describes the measurement of doxepine and desmethyldoxepine in rat plasma. It does not have the above--mentioned disadvantages because it combines the high resolving power of the gas chromatograph with the high sensitivity and specificity of the identification provided by the mass spectrometer. The procedure involves separation by gas-liquid chromatography of doxepine and desmethyldoxepine and detection by mass fragmentography. Concentrations were determined by focusing the mass spectrometer upon the ions at m/e = 277 and m/e = 220 for doxepine, and m/e = 234 for desmethyldoxepine N-acetyl derivative, while promazine was used as internal standard (ion at m/e = 238). Determinations are possible at levels as low as 10 ng/ml of plasma.

With this method, the acute kinetics of doxepine has been determined in the rat. The compound distributes in an apparent V_d (24 L/kg) similar to that of desipramine (21 L/kg). Its plasma levels follow a bi-exponential decay with an α-phase of about 10-12 min, and an apparent plasma half-life of the terminal exponential phase of 54 min. This value is one fourth of that observed for desipramine.

IDENTIFICATION OF PAPAVERINE METABOLITES IN HUMAN URINE BY MASS SPECTROMETRY

M.T. ROSSEEL and F.M. BELPAIRE

J.F. and C. Heymans Institute of Pharmacology, University of Ghent, B-9000 Ghent (Belgium)

SUMMARY

The metabolic pattern of papaverine {1-(3,4-dimethoxybenzyl)-6,7--dimethoxyisoquinoline} in man was studied. Papaverine is excreted in the urine as conjugated metabolites. After hydrolysis of these conjugates, the metabolites were isolated by solvent extraction and thin-layer chromatography. The structure of 5 of these metabolites was elucidated by gas-liquid chromatography with mass spectrometry and by comparison of retention times and mass spectra of reference compounds.

The mass spectra of papaverine and of the 5 metabolites studied were similar. For all compounds, the base peak corresponds to the $(M-H)^+$ ion. In 4 of the 5 metabolites the M^+ ion was 14 a.m.u. smaller than in papaverine; in the 5th metabolite this value was 28 a.m.u. smaller than in papaverine. These results indicate the presence of 4 monodemethylated metabolites and 1 didemethylated metabolite. For each of the metabolites (as for unchanged papaverine) the molecule can be cleaved in two positions; this results in either a CH_2-isoquinoline fragment, or a benzyl fragment; it can then be seen on what moiety of the molecule the hydroxyl group is located. The exact position of the hydroxyl group in the different metabolites is suggested by comparison of the intensity ratios of the $(M-H)^+$ and the $(M-CH_3)^+$ peaks for the unknown products and for the reference compounds, and confirmed by comparison of retention times in gas-liquid chromatography.

After oral administration of papaverine to man, the major metabolites found in the urine were 4'-desmethyl- and 6-desmethylpapaverine; another important metabolite was 3'-desmethylpapaverine; 7-desmethyl- and 4',6--desmethylpapaverine were only present in minor amounts.

INTRODUCTION

Papaverine {1-(3,4-dimethoxybenzyl)-6,7-dimethoxyisoquinoline}, an alkaloid with a 1-benzylisoquinoline structure, is used clinically as a spasmolytic agent. Its metabolism was studied by Axelrod *et al.*[1], and later in our laboratory in more detail, using tritiated papaverine. In the different animal species studied, the drug is predominantly excreted in the bile as conjugated, demethylated metabolites[2-5].

Until now little attention has been paid to the fate of the drug in man. Axelrod *et al.*[1] showed that in man only negligible amounts of unchanged drug are excreted, and they identified the major urinary metabolite as 4'-desmethylpapaverine, but no quantitative data on the metabolism of papaverine in man are available.

Mass spectrometry of papaverine has been used for the analysis of opium alkaloids by Ohashi *et al.*[6], Tatematsu and Goto[7] and Brochmann--Hanssen and Hirai[8]. Recently, by means of mass spectrometry and NMR, we identified four metabolites of papaverine in rat bile[3].

In the present paper the separation and identification of the metabolites of papaverine in the urine of man are described, thin-layer chromatography (TLC), gas-liquid chromatography (GLC) and GLC-mass spectrometry were used. Some preliminary semi-quantitative data are also included.

MATERIALS AND METHODS

Materials

1-(3,4-Dimethoxybenzyl)-6-methoxy-7-hydroxyisoquinoline (metabolite B[+]) was obtained from Professor F. Brochmann-Hanssen, University of California, San Francisco, Calif., U.S.A. 1-(3-Hydroxy-4-methoxybenzyl)--6,7-dimethoxyisoquinoline (palaudine or metabolite A_1) was obtained from B.A. Brossi, Hoffmann-La Roche, U.S.A. 1-(3-Methoxy-4-hydroxybenzyl)-6,7--dimethoxyisoquinoline (metabolite A) and 1-(3,4-dimethoxy)-6-hydroxy--7-methoxyisoquinoline (metabolite C) were purified from rat bile after hydrolysis of conjugates[3]. 1-(3-Methoxy-4-hydroxybenzyl)-6-hydroxy-7--methoxyisoquinoline (metabolite D) was obtained from an extract of rat bile; not enough material could be prepared for crystallization of the substance[3]. 6'-Bromo-papaverine was synthesized for use as internal standard for GLC[9].

[+] We adopted, for simplicity, the terms "metabolites A, A_1, B, C, D" for the different breakdown products described.

Gas-Chrom Q, 60-80 mesh, was obtained from Supelco, (Bellefonte, Pa., U.S.A.) and OV-17 from Applied Science Labs. (State College, Pa., U.S.A).

Glusulase was obtained from Serva (Heidelberg, G.F.R.).

Instrumentation

GLC was carried out on a gas chromatograph, Varian 2100, equipped with a dual flame ionization detector. A glass column (1.83 m x 2 mm I.D.) was packed with 3% OV-17 on 60-80 mesh Gas-Chrom Q. The injection port temperature was 270°, the column temperature was 245° and the detector block temperature was 270°. The hydrogen flow-rate was 30 ml/min and the air flow-rate was 320 ml/min. The nitrogen carrier gas flow was 30 ml/min. The gas chromatograph was connected to a Hewlett Packard 3380A Integrator.

Mass spectra were obtained with a Gnom Matt 111 mass spectrometer, coupled to a Varian 1400-10 gas chromatograph. A glass column (1.5 m x 2 mm I.D.) packed with 3% OV-17 on 60-80 mesh Gas-Chrom Q was used at a temperature of 220° and a helium flow-rate of 10 ml/min. The temperature of the injection port, separator and inlet line was 300°. The electron energy of the ion source was 80 eV. Total emission stream was 270 µA. The ion acceleration voltage was 0.8 KV.

Glassware

All glassware was silanized.

Human experiment

After oral intake of 100 mg papaverine hydrochloride by two volunteers, urine was collected for 18 h.

Procedure

After incubation of 3 ml of urine for 6 h at 37° with glusulase (10,000E) and with acetate buffer (1 *M*, pH 5), Br-papaverine (10 µg) was added. The mixture was adjusted to pH 9 with 10 *M* NaOH and 1 *M* Na_2CO_3 and extracted with two volumes of freshly distilled chloroform. The chloroform was evaporated under nitrogen and the residue was redissolved in 20 µl of chloroform. 1 µl was injected into the gas chromatograph. The rest of the residue (about 19 µl) was spotted on TLC plates (0.25 mm, Silica Gel 60 F--254 on glass, Merck). On the same plates, seven reference products (20 µg) were spotted. The solvent was ethyl acetate-methanol- 25% ammonia (95:5:5). Spots were visualized under UV irradiation.

The areas corresponding to the various metabolites were scraped off and extracted with 4 ml of chloroform; the chloroform extracts were evaporated under nitrogen and the residue was dissolved in 10 µl chloroform; 5 µl were injected into the gas chromatograph-mass spectrometer.

RESULTS AND DISCUSSION

A TLC study of the chloroform extracts of urine samples from volunteers taking papaverine was carried out to obtain preliminary information on the metabolic pathway of papaverine in man. On TLC, six products were separated when the urine had previously been incubated with glusulase. If the urine is not hydrolysed before extraction, only small amounts of metabolites are seen.

A typical chromatogram is shown in Fig. 1, with indication of the R_F values. The R_F values of five of these products are identical with those of the reference substances A, A_1, B, C and D (see Materials). A sixth substance, with an R_F value of 0.17, was only present in very small amount.

To gather more evidence about the structure of these metabolites, and to have a method to quantitate these metabolites a GLC study was done.

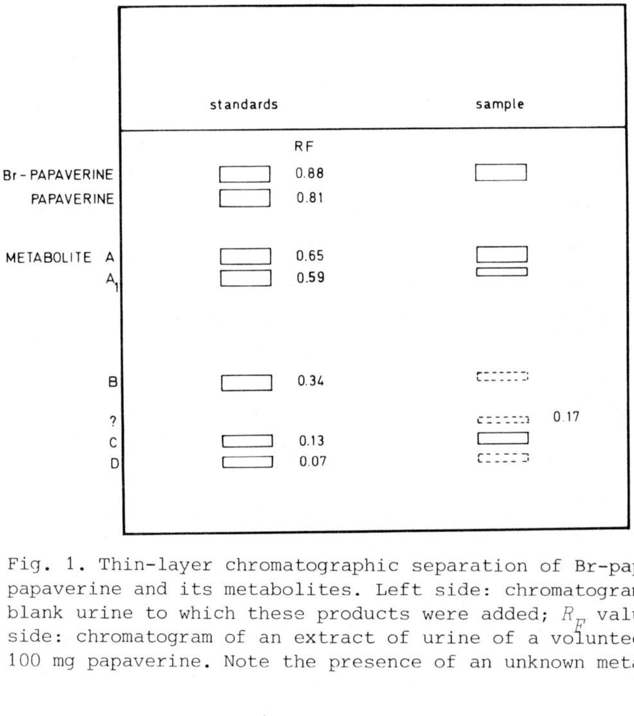

standards sample
RF
Br-PAPAVERINE 0.88
PAPAVERINE 0.81
METABOLITE A 0.65
A_1 0.59
B 0.34
? 0.17
C 0.13
D 0.07

Fig. 1. Thin-layer chromatographic separation of Br-papaverine and of papaverine and its metabolites. Left side: chromatogram of an extract of blank urine to which these products were added; R_F values are given. Right side: chromatogram of an extract of urine of a volunteer after intake of 100 mg papaverine. Note the presence of an unknown metabolite.

A 3% OV-17 column was used because, on it, the five reference compounds (metabolites A, A_1, B, C, D) and Br-papaverine, when added to a blank urine sample, can be separated from each other, as shown in Fig. 2. Papaverine had approximately the same retention time as metabolite A, but no papaverine was present in the urine after administration of papaverine, as shown by TLC.

When urine obtained after intake of papaverine was incubated and extracted, on GLC, six peaks that were not present in the sample of blank urine were seen. The retention times of five peaks, expressed relatively to the retention time of Br-papaverine, are identical with those of the reference compounds. The sixth peak, with a relative retention time of 0.69, corresponds with the unidentified spot on TLC. Fig. 3 shows a typical gas chromatogram.

To confirm the results obtained by GLC and TLC, we performed GLC--mass spectrometric analysis of the different metabolites isolated from the urine. The metabolite with R_F 0.17 for which no reference compound was available, was unstable, and the amount was too small for mass spectrometric study. Figs. 4a and 4b show the mass spectra of the different metabolites in a urine sample after incubation, extraction, TLC and GLC. For comparison, the mass spectrum of papaverine added to urine is also given.

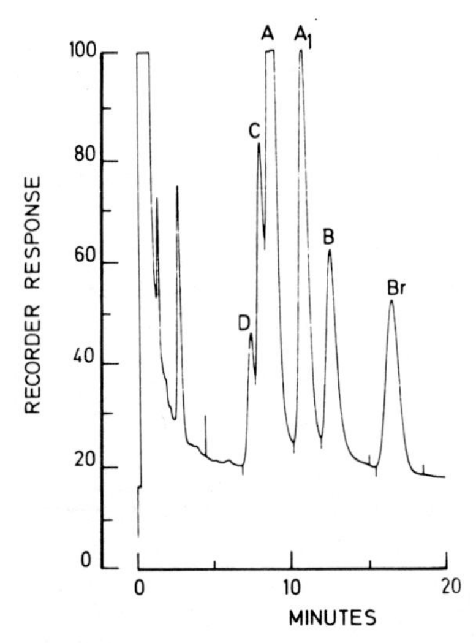

Fig. 2. Gas chromatogram of an extract of blank human urine to which 20 μg of metabolites A, A_1, B, C, an aliquot of metabolite D and 10 μg of Br--papaverine (as internal standard) were added before the extraction procedure.

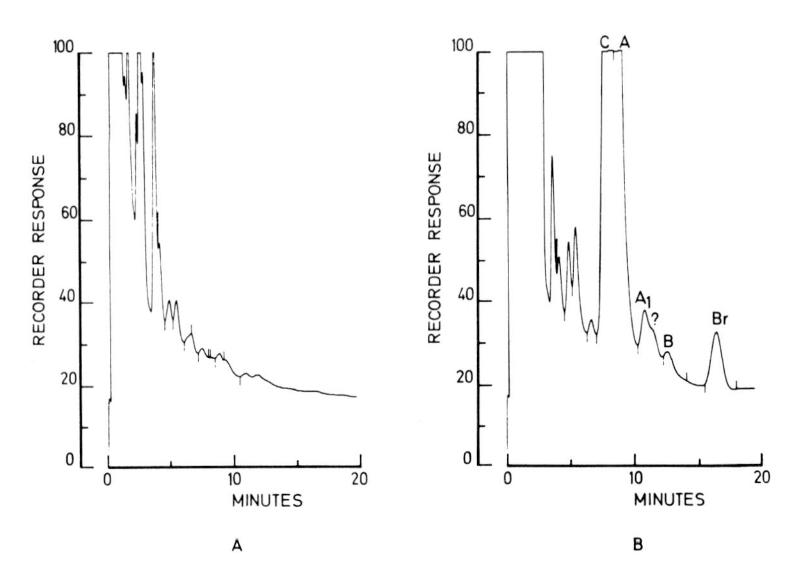

Fig. 3. Gas chromatograms of human urine incubated with glusulase and then extracted. A, Extract from 3 ml blank urine. B, Br-papaverine (10 µg) added to 3 ml of urine of the same subject, obtained 0-2 h after oral intake of 100 mg of papaverine. Note the presence of the metabolites C, A, A_1, B and of the unknown metabolite. The small peak of metabolite D is masked by the large peak of metabolite C.

The mass spectra of the 5 metabolites and of papaverine are similar. The base peak (M-1) of all the substances was due to loss of a hydrogen atom from the molecular ion.

Four of the 5 metabolites had an $M^{+\cdot}$ ion 14 a.m.u. smaller than in papaverine, and the 5th metabolite had an $M^{+\cdot}$ 28 a.m.u. smaller, suggesting that 4 of the metabolites are monodemethylated, and one is didemethylated.

The large peaks at *m/e* 310 $(M-15)^+$ and *m/e* 294 $(M-31)^+$ for the monodemethylated products, and at *m/e* 296 and *m/e* 280 for the didemethylated product, are due to loss of a methyl or a methoxyl radical from one of the aromatic methoxyl substituents; this is typical for aromatic methyl ethers. The usual fragmentation of the $(M-15)^+$ ion followed by loss of carbon monoxide[10] was also observed, as peaks with an *m/e* 282 for the monodemethylated metabolites and with an *m/e* 268 for the didemethylated product. Peaks for $(M-CH_2O)^{+\cdot}$ and $(M-CHO)^+$ are only important for metabolite D.

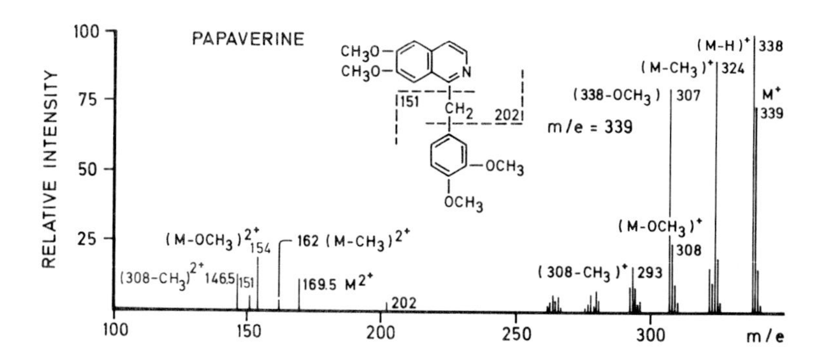

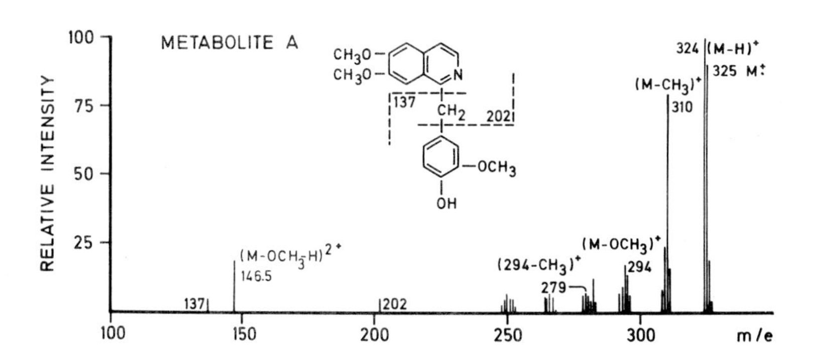

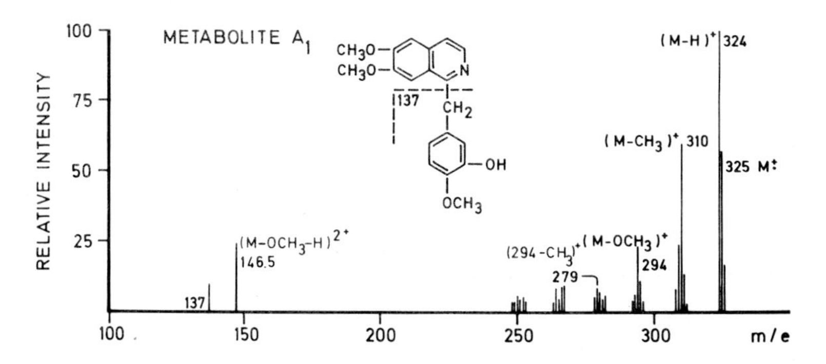

Fig. 4a. Mass spectrum of papaverine added to urine and extracted. Mass spectra of the metabolites A and A_1 isolated by TLC from an extract of human urine incubated with glusulase on combined GLC-mass spectrometry.

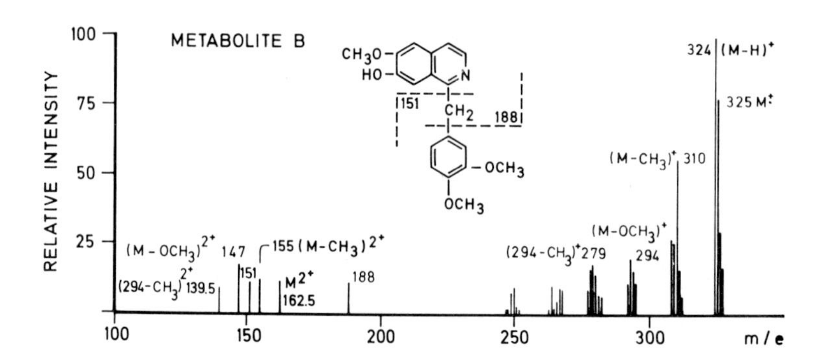

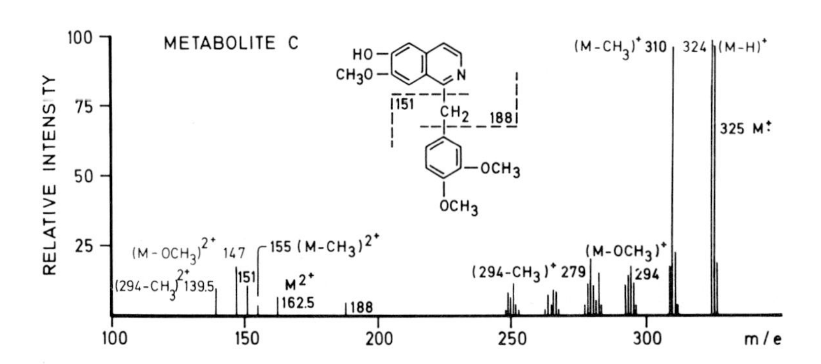

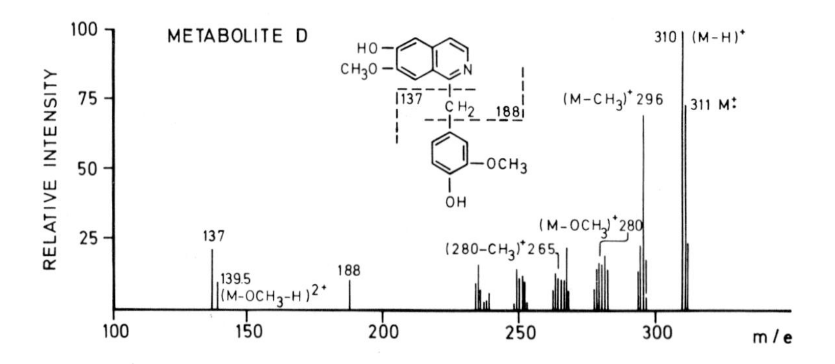

Fig. 4b. Mass spectra of the metabolites B, C and D isolated by TLC from an extract of human urine incubated with glusulase on combined GLC-mass spectrometry.

The molecule is cleaved in two positions as shown in Fig. 5; in this way a CH_2-isoquinoline fragment (fragment I) or a benzyl fragment (fragment II) can be formed. The *m/e* values of these fragment ions in the different metabolites and papaverine are shown in Fig. 5. These peaks are of low intensity but are characteristic. From these values it can be concluded that the hydroxyl group is located on the benzyl moiety or on the isoquinoline moiety of the molecule. Only metabolite A_1 does not show the *m/e* 202 of the CH_2-isoquinoline fragment.

Further fragmentation results in a large number of peaks with very low intensity. Most of the single-charged ions are also present as double--charged ions, but sometimes of very low intensity. Double-charged ions can be expected from these highly stabilized aromatic molecules. The most important double-charged ions are *m/e* 139.5 $(M-OCH_3)^{2+}$, *m/e* 147 $(M-OCH_3)^{2+}$, *m/e* 155 $(M-CH_3)^{2+}$ and *m/e* 162.5 M^{2+} for metabolites B and C; *m/e* 146.5 $(M-OCH_3-H)^{2+}$ for metabolites A and A_1; and *m/e* 139.5 $(M-OCH_3-H)^{2+}$ for metabolite D.

The exact position of the hydroxyl group was suggested by comparison of the intensity ratios of the $(M-H)^+$ and the $(M-CH_3)^+$ peaks for unknown products and reference compounds. These ratios are 1.25 for metabolite A, 1.69 for metabolite A_1, 1.78 for metabolite B and 1.02 for metabolite C. These ratios are the same for the unknown products and the authentic

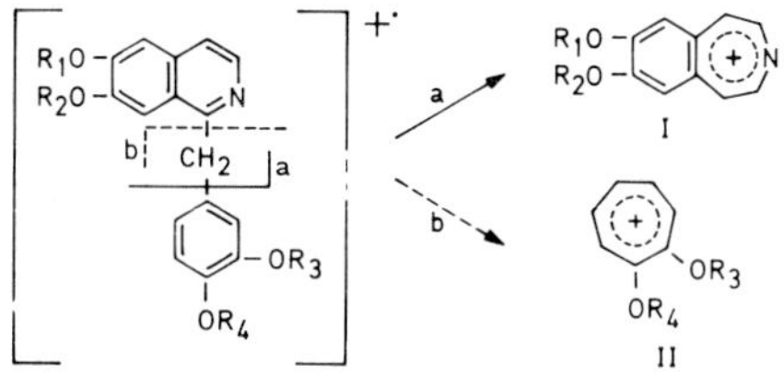

	m/e values	
	I	II
PAPAVERINE	202	151
METABOLITE A	202	137
METABOLITE A_1	–	137
METABOLITE B	188	151
METABOLITE C	188	151
METABOLITE D	188	137

Fig. 5. Benzyl fragment (II) and CH_3-isoquinoline fragment (I) from benzylisoquinolines on mass spectrometry, for papaverine and its different metabolites.

standards, when studied under the same GLC-mass spectrometric conditions.

In conclusion it can be said that the GLC-mass spectrometric analysis confirms that papaverine is metabolized to at least four conjugated monodemethylated metabolites and one didemethylated compound, as shown in Fig. 6. This corresponds approximately with what has been found in the bile of different animal species except that metabolite A_1 has not been detected in animals[4].

Furthermore, from TLC and GLC data it can be concluded that the conjugates of metabolites A and C are the major products excreted in human urine, and they account for about 40% of the given dose. Metabolites A_1 and D are excreted in smaller amounts, and metabolites B and a non--identified substance are only present in small amounts.

ACKNOWLEDGEMENTS

The authors thank Professor Dr. M. Bogaert, Dr. D. De Keukeleire and Dr. C. Van De Sande for discussing the manuscript. This work was supported be the Medical Research Fund, Belgium.

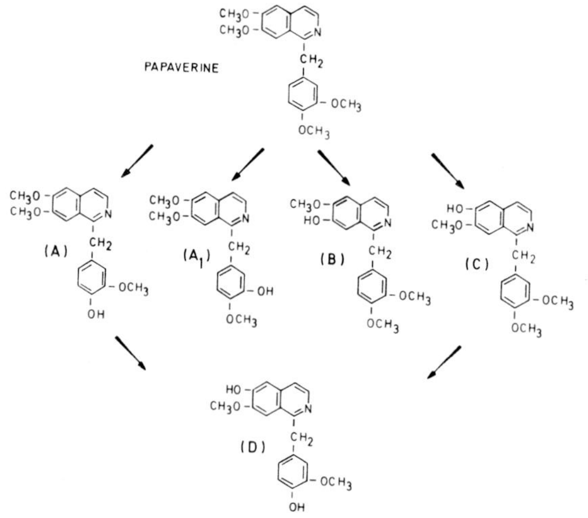

Fig. 6. Metabolic pathway of papaverine in man.

REFERENCES

1 J. Axelrod, R. Shofer, J.K. Inscoe, W.M. King and A. Sjoerdsma, *J. Pharmacol. Exp. Ther.*, 124 (1958) 9.
2 F.M. Belpaire and M.G. Bogaert, *Biochem. Pharmacol.*, 22 (1973) 59.
3 F.M. Belpaire, M.G. Bogaert, M.T. Rosseel and M. Anteunis, *Xenobiotica*, 5 (1975) 413.
4 F.M. Belpaire and M.G. Bogaert, *Xenobiotica*, 5 (1975) 421.
5 F.M. Belpaire and M.G. Bogaert, *Xenobiotica*, 5 (1975) 431.
6 M. Ohashi, J.M. Wilson, H. Budzikiewicz, M. Shamma, W.A. Slusarchyk and C. Djerassi, *J. Amer. Chem. Soc.*, 85 (1963) 2807.
7 A. Tatematsu and T. Goto, *Yakugaku Zasshi*, 85 (1965) 152.
8 E. Brochmann-Hanssen and K. Hirai, *J. Pharm. Sci.*, 57 (1968) 940.
9 T. Anderson, *Justus Liebigs Ann. Chem.*, 94, 235.
10 H. Budzikiewicz, C. Djerassi and D.H. Williams, *Mass Spectrometry of Organic Compounds*, Holden-Day, San Francisco, Calif., 1967, p. 237.

DIRECT GAS CHROMATOGRAPHIC-MASS SPECTROMETRIC MEASUREMENT OF PROSTAGLANDINS FROM BIOLOGICAL SOURCES

R.W. KELLY

Medical Research Council, Unit of Reproductive Biology, 2 Forrest Road, Edinburgh EH1 2QW (Great Britain)

SUMMARY

One of the particular advantages of gas chromatography-mass spectrometry (GC-MS) analysis is the high specificity; this allows sensitive measurements in dirty mixtures as well as giving reliable confirmation of results from quicker and cheaper techniques such as radioimmunoassay.

I have developed the use of *tert*.-butyl dimethylsilyl ethers and oximes to give an analytical technique that allows the measurement of subnanogram quantities of prostaglandins (of both the E and F series) in one derivatized sample.

The technique involves oximation of the ketonic prostaglandins in aqueous solution, followed by extraction and derivatization as methyl ester *tert*.-butyl dimethylsilyl ethers. The strong M-57 ions produced from these derivatives (*m/e* 653 from $PGF_{2\alpha}$ and *m/e* 666 from PGE_2) allow the simultaneous measurement of a variety of E and F compounds in one GC run. An added advantage of using *tert*.-butyl dimethylsilyl ethers is that several internal standards can be used.

The use of this technique in the measurement of prostaglandins in tissue and semen is described.

INTRODUCTION

Although most measurements of steroids and prostaglandins are now done by radioimmunoassay (RIA), this technique has limitations, particulary in measurements in biological fluids from new sources. Gas chromatography-mass spectrometry (GC-MS) is a specific technique which can be used with relatively crude extracts, and being a physico-chemical technique, can be applied quickly to new compounds. Thus GC-MS complements RIA in many instances. The use of new derivatives, particularly *tert*.-butyl dimethylsilyl (tBDMS) ethers, in quantitative GC-MS analysis increases both the specificity and the usefulness of the technique. This paper describes the use of tBDMS derivatives in quantitative analyses of prostaglandins and gives two examples; one where RIA required confirmation and one where RIA was impossible because the compound to be measured had only recently been discovered.

A method for the preparation of tBDMS ethers using *tert*.-butyl dimethylsilyl chloride and imidazole was reported by Corey and Venkateswarlu in 1972[1] in a paper describing the use of this derivative in synthesis. Since then several papers have described the use of this derivative for the analysis of steroids[2-4] and prostaglandins[5-7]. The preparation of tBDMS ethers has been recently improved by the use of Sephadex^{R} LH-20 columns to purify the product[4,7].

For the present study, a DuPont 490B mass spectrometer was coupled through a glass jet separator to a Varian 1400 gas chromatograph fitted with a 10- or 20-m glass SCOT column (SGE Ltd.) coated with SE-30. The injection system consisted of a single (1/8 in. O.D., 1/16 in. I.D.) glass-lined steel tube (SGE Ltd.) with no sample split. A typical column temperature was 265° with a helium carrier gas flow-rate of 5 ml/min. Under these conditions tBDMS prostaglandins eluted in about 5 min. Although the resolution of the capillary column was compromised by the relatively high flow-rates, reasonable resolution was available with relatively short retention times. A further advantage of the capillary GC-MS was that the smaller surface area of the support gave less degradation of trace amounts of compound. Multiple--ion detection (MID) was achieved by synchronously switching the accelerating voltage and the output from the signal amplifier. The signals from the tuned ions were fed to a Rikadenki 6-channel pen recorder. Previously described methods were used for the preparation of tBDMS ethers[7], oximes[7] and butyl boronates[8].

ANALYSIS OF E AND F PROSTAGLANDINS

The mass spectra of the oxime/tBDMS derivatives of prostaglandins E_1 and E_2 show strong base peaks at *m/e* 668 (E_1) and *m/e* 666 (E_2) although the oxime tBDMS has a longer retention time than does the methyl oxime tBDMS, the M-57 peak in the oxime tBDMS is considerably stronger than in the methyl oxime tBDMS and the derivative is less affected by thermal degradation. The mass spectrum of $F_{2\alpha}$ tris-(tBDMS ether) methyl ester is liable to modification by thermal degradation, as we have previously reported[7]. However, the DuPont 490B with high speed pumping and glass jet separator routinely gives a mass spectrum with minimal contribution from thermal processes.

Because of the almost universal nature of the M-57 peak in the spectra of tBDMS prostaglandins, modification to the molecule, although it may alter the mass of the M-57 peak, did not appreciably affect the main fragmentation pathways. Thus, apart from the 3,3,4,4-tetradeutero prostaglandins, various other modified prostaglandins may be used as internal standards. I have used 20-ethyl prostaglandin $F_{2\alpha}$ (a gift from Dr. N.S. Crossley of I.C.I. Pharmaceuticals Ltd.) as a satisfactory internal standard for the measurement of prostaglandin $F_{2\alpha}$. Because of an appreciable (1%) contribution to the 653 channel ($F_{2\alpha}$) from the 3,3,4,4-tetradeutero $F_{2\alpha}$, this modification improved the sensitivity of the analysis. In the analyses of E prostaglandins we have used the 3,3,4,4-tetradeutero compounds (a gift from Dr. U. Axen of the Upjohn Co.) as the internal standard.

In a recent study of samples from endometrial biopsy (H.Maathuis, unpublished results), the $PGF_{2\alpha}$ and PGE_2 levels were analysed. The tissue samples were homogenized in ethanol and centrifuged; the supernatant was stored at -20° until analysed. To measure Es and Fs in 1 ml of the ethanol extract, standards (3,3,4,4-tetradeutero E_2 and 20 ethyl $F_{2\alpha}$ or 3,3,4,4--tetradeutero $F_{2\alpha}$) were first added, then the ethanol was transferred to a tube containing 10 ml of 1% hydroxylamine hydrochloride in pyridinium acetate (1.5 *M*, pH 5) and the oxime/methyl ester/tBDMS derivative was formed. The resulting derivatized extract is suitable for the measurement of E prostaglandins (E_1 at *m/e* 668 and E_2 at *m/e* 666) and F prostaglandins ($F_{1\alpha}$ at *m/e* 655 and $F_{2\alpha}$ at *m/e* 653) in one GC run, the only limitation being the number of channels on the recorder that can be used conveniently. Fig. 1 shows a typical trace of the analysis of 1 ng of E_2 and 1 ng of $F_{2\alpha}$.

The precision of GC-MS analysis determined by the repetitive injection of 1 ng of $PGF_{2\alpha}$ derivatized with 20-ethyl $F_{2\alpha}$ as internal standard, was 9% (S.D., n = 12). The inter-batch precision obtained from duplicates analysed at different times was 23.5% for $F_{2\alpha}$ (n = 10 pairs) and 24% for E_2 (n = 11 pairs). This relatively poor precision is a result of the wide range of prostaglandin levels found in endometrial tissue. To ensure that there was no agent in the extracts that interfered with the derivatization, different volumes of an extract were diluted to 2 ml with ethanol and analysed for $F_{2\alpha}$ concentration. The results, which show no systematic

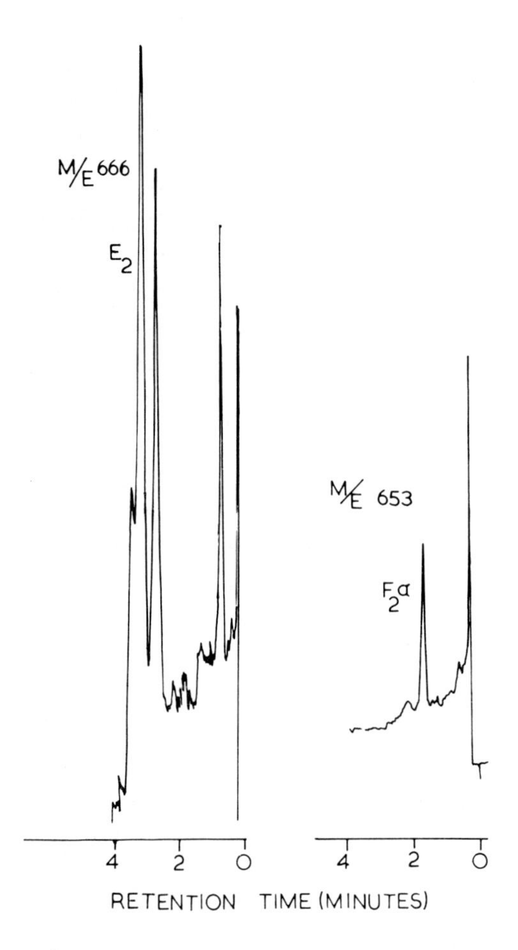

Fig. 1. Tuned-ion traces of PGE_2 and $PGF_{2\alpha}$ from the analyses of endometrial tissue. Two oxime isomers are visible in the E_2 trace; the first is used for quantitation. Traces of the internal standard are omitted for the sake of clarity.

variation, are shown in Table I.

Checks on the accuracy of the method were made by comparing levels of $F_{2\alpha}$ using methyl ester/tBDMS ethers with 3,3,4,4-tetradeutero $F_{2\alpha}$ and 20-ethyl $F_{2\alpha}$ as internal standards with measurements made on methyl ester/butyl boronate/trimethyl silyl (TMS) derivatives with tetradeutero $F_{2\alpha}$ as the internal standard. Analyses using this last combination have already been characterized[8]. The results, in Table II, show that the three techniques compare favourably.

Table I

Analysis of different volumes of ethanol extract

All extracts were made up to 2 ml before analysis

Volume of ethanol extract	Measured $F_{2\alpha}$ (ng/g of tissue)
0.25	96
0.5	82
1	80
2	111

Table II

Total $F_{2\alpha}$ content of tissue (ng) assayed with different derivatives

combination	a	b	c
Derivative	methyl ester- -tris-(tBDMS ether)	methyl ester- -tris-(tBDMS ether)	methyl ester- -butyl boronate- -bis-(tBDMS ether)
Internal standard	tetradeutero $F_{2\alpha}$	20-ethyl $F_{2\alpha}$	tetradeutero $F_{2\alpha}$
	141	77	111
	168	96	143
	276	243	350
	150	147	143
	99	98	207
	488	404	413

Correlation coefficients: a:b = 0.97, a:c = 0.88, b:c = 0.92.

MEASUREMENT OF 19-HYDROXY PROSTAGLANDIN $F_{2\alpha}$

It has been claimed that there are thirteen prostaglandins in human semen[9]. More recently the finding of 19-hydroxy E prostaglandins[10,11] and the virtual absence of A, B and 19-hydroxy A and B prostaglandins previously claimed as major constituents, suggest that the dehydrated prostaglandins

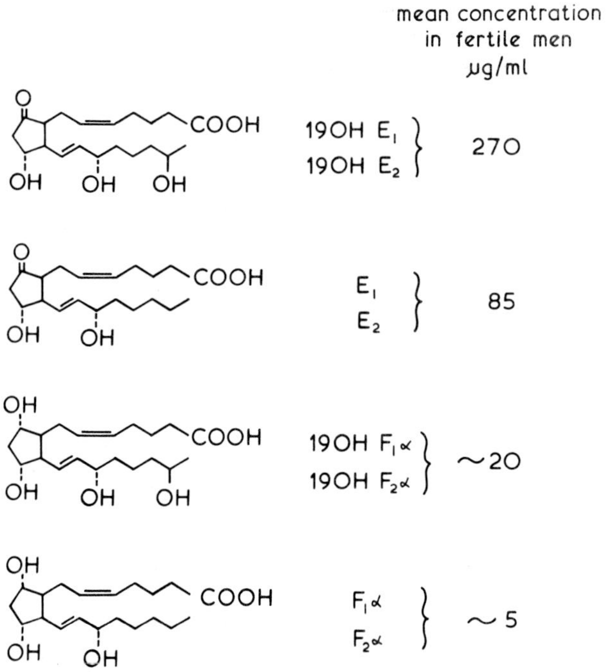

mean concentration in fertile men μg/ml

COOH, OH OH OH	19OH E_1, 19OH E_2	270
COOH, OH OH	E_1, E_2	85
OH, COOH, OH OH OH	19OH $F_{1\alpha}$, 19OH $F_{2\alpha}$	~20
OH, COOH, OH OH	$F_{1\alpha}$, $F_{2\alpha}$	~5

Fig. 2. Prostaglandins in human semen.

are artifacts. Moreover, 19-hydroxy F prostaglandins were found in human semen at concentrations of around 20 μg/ml[12]. The main prostaglandins of human semen as now known are shown in Fig. 2.

Several workers have shown a correlation between otherwise unexplained infertility and concentration of prostaglandin E[9,13,14]. To ascertain the relevance of the newly discovered PGs to infertility I wished to analyse a complete spectrum of prostaglandins in semen. Although it has been possible to use GC alone for the measurement of 19-hydroxy E prostaglandins[15], the lower levels of 19-hydroxy Fs require a more sensitive and selective detector. Accordingly, a GC-MS technique was used for their measurement.

Because of the high mass of the base peak of the methyl ester tetra--tBDMS derivative of 19-hydroxy $F_{2\alpha}$ (*m/e* 785) and the corresponding long GC retention times, the methyl ester/butyl boronate/tBDMS ether was used as derivative. This derivative, which can be made by the modified procedure of tBDMS preparation, has a base peak at *m/e* 621 for 19-OH $PGF_{2\alpha}$. The mass spectrum of this derivative is shown in Fig. 3. This mass spectrum is subject to modification by thermal degradation; the first time the spectrum

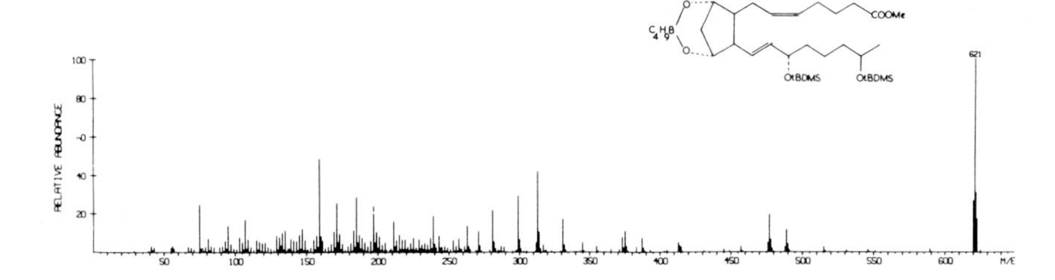

Fig. 3. Mass spectrum of methyl ester butyl boronate tBDMS ether of 19-OH $PGF_{2\alpha}$.

was recorded (on an A.E.I. MS12 with Watson-Bieman separator) no silanizing agent had been through the system for several days and the base peak was at *m/e* 75 with M-57 as 60% of the base peak. However, after the system has been silanized the spectrum in Fig. 3 was obtained.

For the measurement of 19-hydroxy $PGF_{2\alpha}$ in semen it is sufficient to add ethanol to the sample of semen, centrifuge, and evaporate before derivatization. The ethyl ester of synthetic 19-hydroxy $F_{2\alpha}$ (a gift from Dr. N.S. Crossly), prepared by using diazoethane, was introduced as the internal standard after the sample has been methylated. Although this technique does not allow any correction for poor methylation, this stage is almost invariably quantitative.

The detection limit of the GC-MS assay is less than 200 pg of 19-hydroxy $F_{2\alpha}$ injected into the mass spectrometer. The precision for repeated injections of 100 ng of 19-hydroxy $PGF_{2\alpha}$ from derivatized semen was 3.9% (S.D., n = 12) and for the repeated derivatization and assay of an ethanolic extract of semen the precision was 11.7% (S.D., n = 8). In a sample of semen having a normal concentration of 19-hydroxy $F_{2\alpha}$, approximately 100 ng of this compound would be injected into the gas chromatograph-mass spectrometer for measurement. The GC tuned-ion trace of the *m/e* 621 ion from a typical analysis is shown in Fig. 4.

The above analytical schemes demonstrate the versatility of tBDMS derivatives in quantitative analysis by MID techniques. The common nature of the M-57 base peak, the stability of the derivative and the high mass of the ion followed will probably make tBDMS ethers the derivatives of choice in future quantitative analyses not only of prostaglandins but also of a wide range of other compounds.

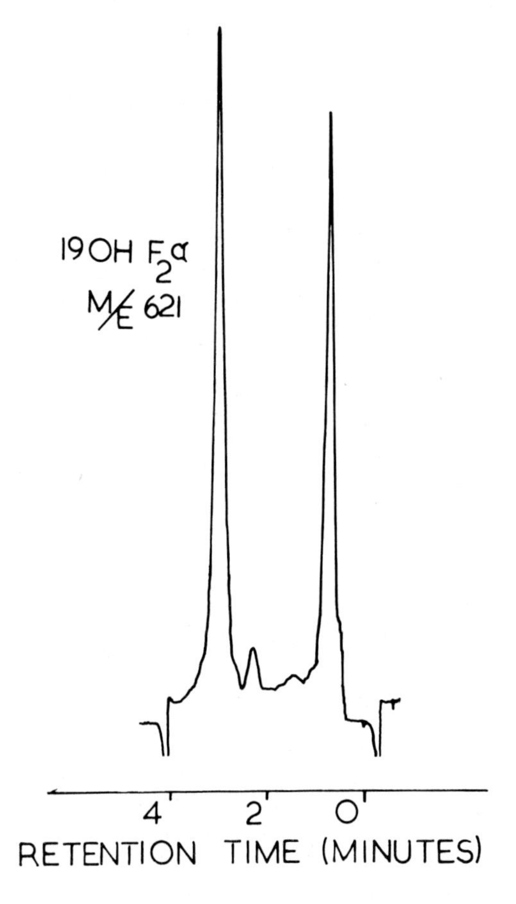

Fig. 4. A typical tuned-ion trace from the analysis of 19-OH $F_{2\alpha}$ from human semen.

ACKNOWLEDGEMENTS

I would like to thank Professor R.V. Short for his helpful advice and encouragement. I am grateful to Dr. N.S. Crossley of I.C.I. Pharmaceuticals Ltd., and Dr. J.E. Pike and Dr. U. Axen of the Upjohn Co. for the supply of prostaglandins; to P.L. Taylor for running the mass spectrum of 19-OH $F_{2\alpha}$ and to Irene Cooper for technical assistance.

REFERENCES

1 E.J. Corey and A. Venkateswarlu, *J. Amer. Chem. Soc.*, 94 (1972) 3190.
2 D.S. Millington, *J. Steroid Biochem.*, 6 (1975) 239.
3 G. Phillipou, D.A. Bigham and R.F. Seamark, *Steroids*, 26 (1975) 516.
4 S.J. Gaskell and C.J.W. Brooks, *Biochem. Soc. Trans.*, 4 (1976) 111.
5 R.W. Kelly and P.L. Taylor, in A. Frigerio and N. Castagnoli (Editors), *Advances in Mass Spectrometry in Biochemistry and Medicine*, Vol. 1, Spectrum, New York, 1976, p. 449.
6 J.T. Watson and B. Sweetman, *J. Org. Mass Spectrom.*, 9 (1974) 39.
7 R.W. Kelly and P.L. Taylor, *Anal. Chem.*, 48 (1976) 465.
8 R.W. Kelly, *Anal. Chem.*, 45 (1973) 2079.
9 M. Bygdeman, B. Fredricsson, K. Svanborg and B. Samuelsson, *Fertil. Steril.*, 21 (1970) 622.
10 P.L. Taylor and R.W. Kelly, *Nature (London)*, 250 (1974) 665.
11 H.T. Jonsson, B.S. Middleditch and D.M. Desiderio, *Science*, 187 (1975) 1093.
12 P.L. Taylor and R.W. Kelly, *FEBS Lett.*, 57 (1975) 22.
13 D.F. Hawkins, in P.W. Ramwel and J. Shaw (Editors), *Prostaglandin Symposium of the Worcester Foundation*, Interscience, New York, 1968, p. 1.
14 J.G. Collier, R.J. Flower and S.L. Stanton, *Fertil. Steril.*, 26 (1975) 868.
15 I. Cooper and R.W. Kelly, *Prostaglandins*, 10 (1975) 507.

URINARY HYDANTOIN-5-PROPIONIC ACID MEASURED BY SELECTED ION MONITORING, AN INDICATOR OF FOLIC ACID DEFICIENCY AND FORMIMINOTRANSFERASE DEFICIENCY

A. NIEDERWIESER, A. MATASOVIC and B. KEMPKEN

Medizinisch-Chemische Abteilung, Universitäts-Kinderklinik, 8032 Zurich (Switzerland)

A method for the quantitation of urinary hydantoin-5-propionic acid (HPA) was developed, using the high specificity of selected ion monitoring (SIM). Permethylation of urine is performed by methyl iodide and tetramethylammonium hydroxide in methanol-dimethylformamide with α,α,γ--trideuterated HPA as an internal standard. The sample is separated by gas chromatography on a 2 m x 2 mm column of 0.5% Carbowax 20M and Chromosorb AW DMCS at 190° within less than 6 min and monitored continuously at *m/e* 214 (HPA) and 216 + 217 (internal standard) at 15 eV. The high specificity of SIM allows us to work quickly without sample precleaning. Using a non-specific detector, reliable quantitation would not be possible without excessive pre-purification of the sample.

HPA, a metabolic end-product of histidine, is formed by oxidation of 4-imidazolone-5-propionate (ImOPA) with aldehyde oxidase. ImOPA, an unstable key metabolite, is also hydrolyzed to formiminoglutamate (FIGLU), and FIGLU is transformed into glutamate by a tetrahydrofolate-dependent formiminotransferase.

In normal adults, 4.5 ± 2.2 (n = 24) and 46.0 ± 16.4 mmole HPA per mole creatine (n = 17) were found before and after ingestion of free histidine, respectively (three doses of 66 mg His/kg each, with 4-h intervals). The variation coefficient was 5.6%, when measurements were based on *m/e* 217 only, and less than 1.5% when based on the sum of *m/e* 217 + 216. Recovery was higher than 96%.

In general or functional folic acid deficiency, excretion of FIGLU and HPA is increased after ingestion of histidine. The correlation curve between FIGLU and HPA excretions suggests that HPA measurement may be a more sensitive tool to detect folic acid deficiency. In two cases of formiminotransferase deficiency we found massive amounts of urinary HPA in addition to FIGLU, even without histidine loading. Measurement of HPA also seems to be of value in other defects of histidine catabolism.

Details will be published in *Clin. Chim. Acta*, and *Ped. Res.*, 10 (1976) 215.

PICOMOLE DETERMINATION OF COENZYME B_{12} BY MASS FRAGMENTOGRAPHY

B. ZAGALAK, U. REDWEIK, BIANKA KEMPKEN and H.-CH. CURTIUS

University of Zurich, Institute of Pediatrics, Medical Chemical Division, Steinwiesstrasse 75, 8032 Zurich (Switzerland)

SUMMARY

The principle of the determination is based on photolytic cyclization of the 5'-deoxyadenosyl part of coenzyme B_{12} into 8,5'-cyclic adenosine, which is finally analyzed after full silylation in a coupled gas chromatography-multiple ion detection system. Deuterated coenzyme B_{12} was used as an internal standard. The isolation of coenzyme B_{12} from biological material and purification is described. The electron-impact mass spectra of $(TMS)_3$-8,5'-cyclic-adenosine and $(TMS)_3$-8,5'-cyclic-adenosine-5',5'-d_2 as well as the electron-impact fragmentation pattern are presented.

Mass fragmentography was also successfully applied in the determination of the deuterium content in the coenzyme B_{12} derived from enzymic reactions.

INTRODUCTION

Coenzyme B_{12} (5'-deoxyadenosylcobalamin) participates in several enzymic reactions as a hydrogen carrier. In general, the coenzyme occurs in biological material in catalytic amounts. Therefore, known chemical and spectroscopic methods for the determination of coenzyme B_{12} are useless. On the other hand, coenzyme B_{12} can be determined at catalytic level by any enzymic, coenzyme B_{12}-dependent reaction[1,2], but unfortunately, these systems are extremely sensitive to other cobalamins and some nucleotides. These compounds strongly inhibit the coenzyme B_{12}-dependent enzymic reactions. Any valuable coenzyme B_{12} assay requires complicated and time--consuming purification and chromatographic steps.

Earlier data on photolysis of coenzyme B_{12} showed the formation of

8,5'-cyclic-adenosine, adenosine-5'-aldehyde and other adenine nucleosides [3-6]. The photolytic cyclization of the 5'-deoxyadenosyl remainder of the coenzyme B_{12} into 8,5'-cyclic-adenosine seems to be typical for the homolytic cleavage of the sigma 5'-carbon-cobalt in coenzyme B_{12} [4,7]. The formation of this unusual 8,5'-cyclic-adenosine is the principle of the method for the determination of coenzyme B_{12} presented here. This includes the use of coenzyme B_{12}-5',5'-d_2 as an internal standard.

RESULTS

Anaerobic photolysis (Fig. 1) of aqueous solutions of highly deuterated coenzyme B_{12} at C-5' shows that the cleavage of the carbon-cobalt bond and the cyclization into 8,5'-cyclic-adenosine proceed practically without loss of deuterium. This observation, and the easy availability of

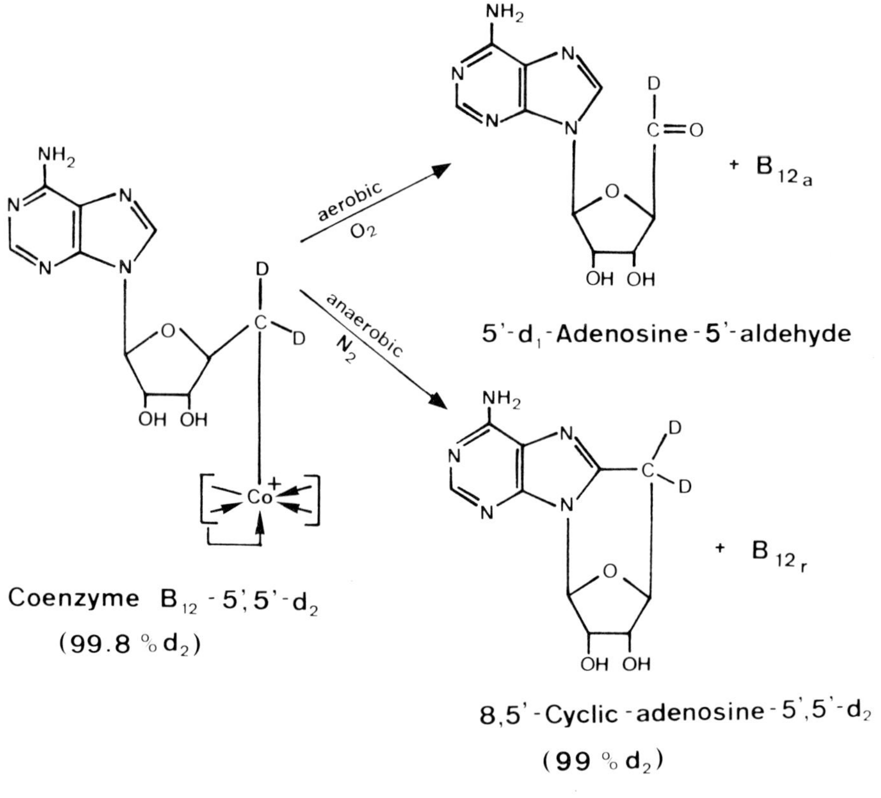

Fig. 1. Photolysis of deuterated coenzyme B_{12}. Experimental conditions: 4×10^{-5} *M* aqueous solutions, 25°, 30 min bright daylight. Details are given in the experimental part.

labeled coenzyme B_{12}, enabled us to use coenzyme B_{12} deuterated at C-5' in the sugar part as an ideal internal standard in our quantitative mass spectrometry coupled with gas chromatography.

The idea of the determination of coenzyme B_{12} as 8,5'-cyclic-adenosine receives further support from a very favorable electron impact fragmentation pattern of its tris(trimethylsilyl) derivative (Fig. 3), resulting in a high abundance of the following major ions: *m/e* 465, M^{+} (42% rel.int.); *m/e* 450 (M^{+}-CH_3) (21%) and *m/e* 221 (base+1H)-CH_3 (100%). This fragmentation is similar to the fragmentation pattern of trimethylsilyl or trifluoroacetyl derivatives of adenine nucleosides[8-10], except for the high abundance of the M^{+}. The base fragment itself (adenine) *m/e* 206 and the

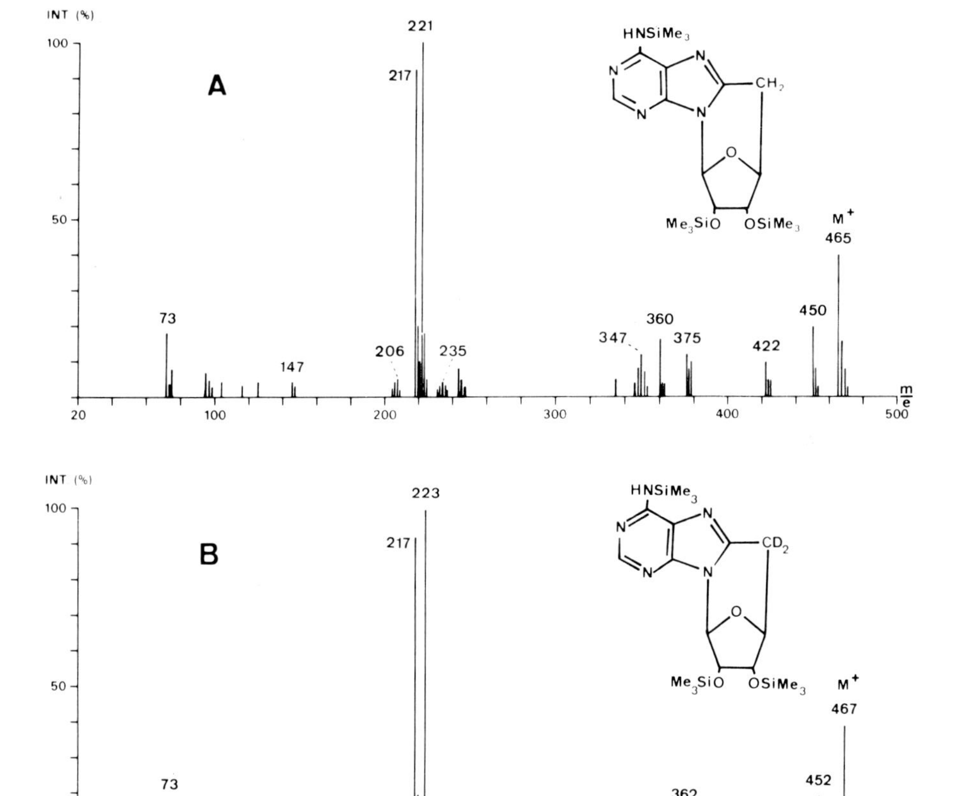

Fig. 2. Electron-impact mass spectra of: A, tris(trimethylsilyl)-8,5'-cyclic-adenosine; B, tris(trimethylsilyl)-8,5'-cyclic-adenosine-5',5'-d_2.

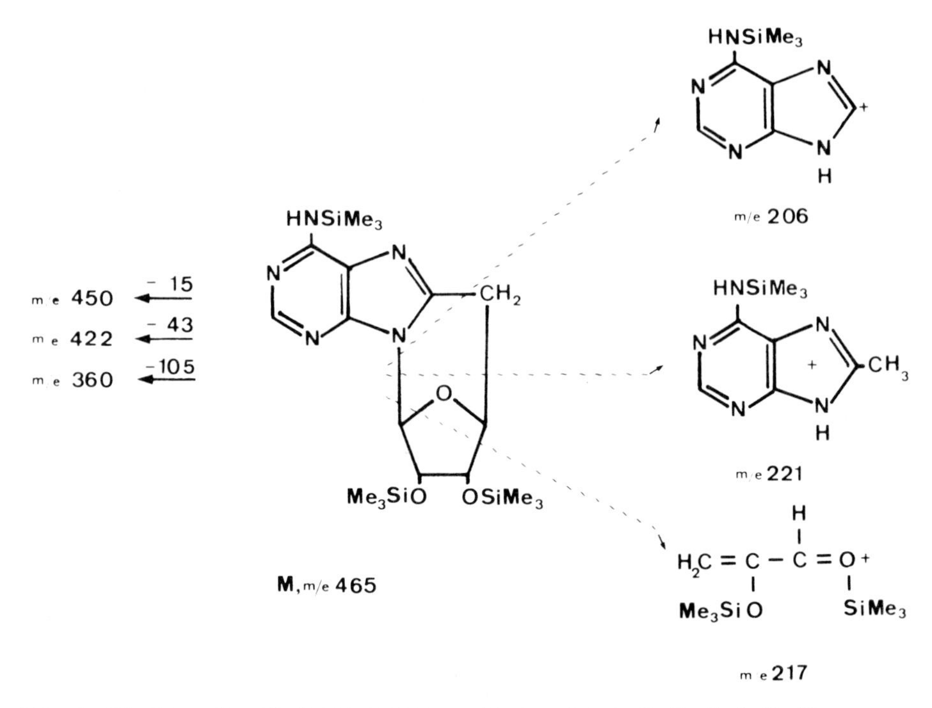

Fig. 3. Electron impact fragmentation pattern of tris(trimethylsilyl)--8,5'-cyclic-adenosine.

rearranged species base+1H and base+2H are at low abundance. The other observed fragment *m/e* 217 derives from the ribose skeleton and corresponds to either $C_{1'}$-$C_{2'}$-$C_{3'}$ or $C_{2'}$-$C_{3'}$-$C_{4'}$. Ions of mass 73, 147 and 217 are frequently observed in the mass spectra of trimethylsilylated compounds; therefore their presence has no significant importance.

To obtain relatively pure samples of coenzyme B_{12} from biological material, we extracted corrinoids with trichloroacetic acid and selectively adsorbed them on Amberlite XAD-2 (ref. 11). The overall yield of the extraction and purification of corrinoids varied between 90 and 100%. The traditional methods used for specific purification of corrinoids, namely adsorption on charcoal or phenol extraction were not efficient enough.

Assay of coenzyme B_{12}

The sample (0.1-0.5 g of liver) was homogenized for 10 min with an excess of aqueous 10% trichloroacetic acid and centrifuged. The supernatant liquid was collected and the precipitate extracted twice again in the same manner. The combined extracts (protected from light) were passed through an

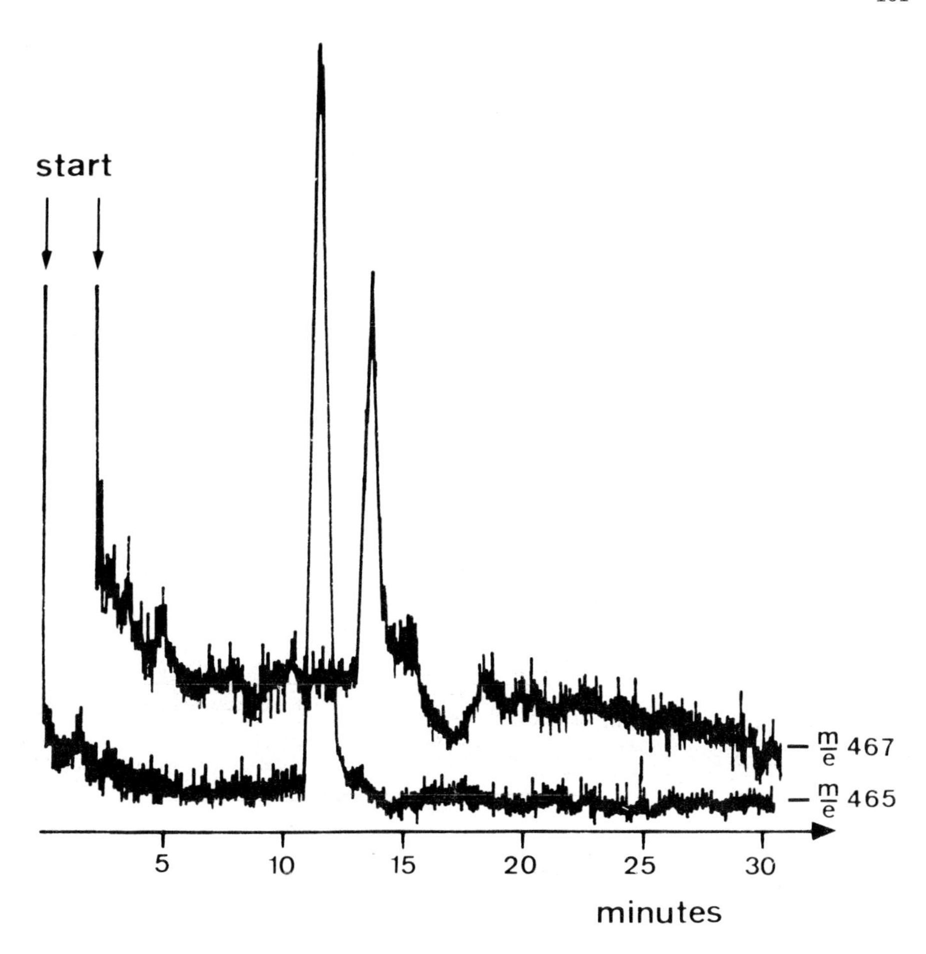

Fig. 4. Mass fragmentogram of tris(trimethylsilyl)-8,5'-cyclic-adenosine and tris(trimethylsilyl)8,5'-cyclic-adenosine-5',5'-d_2 derived from anaerobic photolysis of coenzyme B_{12} and coenzyme B_{12}-5',5'-d_2 (internal standard), 5.8 and 3.5 pmole/µl, respectively.

Amberlite XAD-2 column (8 mm x 40 mm, 50-80 µm; Serva, Heidelberg, G.F.R.) and desalted by washing with water. The coenzyme B_{12} was eluted from the column with 10 ml of the elution mixture (water saturated with ethyl acetate + 4% of *tert*.-butanol + 1% of 30% aqueous ammonia, v/v). The eluate was evaporated to dryness and the residue taken up with 3 ml of water, mixed with coenzyme B_{12}-5',5'-d_2 as an internal standard (which can be added before the extraction) and flushed with oxygen-free nitrogen for 45 min. The sample was then exposed to bright daylight for 30 min (or for 60 min, 100-W tungsten lamp at a distance of 10 cm, room temperature). After the completion of the photolysis the solvent was evaporated and the sample dried azeotropically with dichloromethane in a rotary evaporator. The dried sample was silylated with 200 µl of the silylating mixture (pyridine-trimethylchlorosilane-hexamethyldisilazane, 2:1:1, v/v/v) for 60 min at room temperature and evaporated to dryness. Finally, the sample was dissolved in 50 µl of bis(trimethylsilyl)-trifluoroacetamide (BSTFA), and 0.5-1.0 µl was directly analyzed in a coupled GC-MID system by monitoring previously selected and focused masses which correspond to ions derived from the d_0- and d_2-cyclic-adenosine-$(TMS)_3$: *m/e* 465 and 467 or *m/e* 450 and 452 or *m/e* 221 and 223.

This method enabled us to determine coenzyme B_{12} in pmole quantities (Fig. 5) when the typical *m/e* 465 and 467 peaks were recorded or below pmole quantities when the less characteristic *m/e* 221 and 223 peaks were monitored.

Previous kinetic data on the function of coenzyme B_{12} as a hydrogen carrier suggest that the dioldehydrase-coenzyme B_{12} complex does not dissociate at an appreciable rate relative to the rates of the catalytic

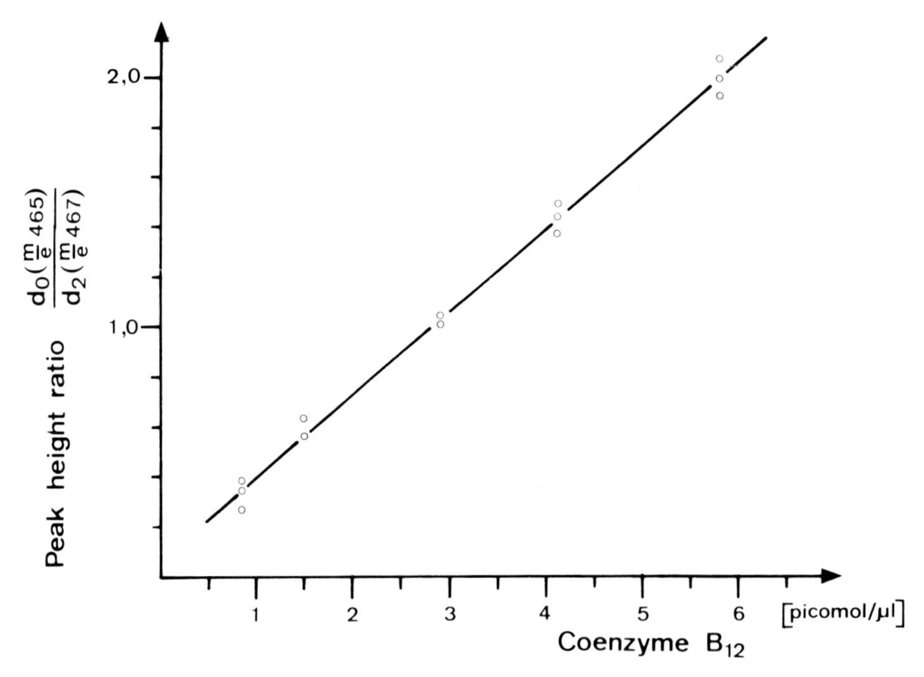

Fig. 5. Standard curve for the quantitative determination of coenzyme B_{12}. The standard curve was prepared by analyzing standard solutions of coenzyme B_{12} and coenzyme B_{12}-5',5'-d_2.

reaction[7,12]. However, the existence and importance of this dissociation cannot be excluded in the interpretation of kinetic data. Investigation of this problem could shed more light on the mechanism of the hydrogen transfer process. Recent data were obtained from kinetic studies performed mainly with tritiated substrates of coenzyme B_{12}. The existence of a large isotope effect in the tritium transfer in this system complicates the explanation of the kinetic results and hence the interpretation of the mechanism of the hydrogen transfer process. To simplify this problem, we replaced tritium and hydrogen with deuterium and used almost completely deuterated substrates in our investigations carried out with glyceroldehydratase, another coenzyme B_{12}-dependent enzyme. In these experiments the coenzyme B_{12} was extracted from enzymic mixtures, purified, and derivatives were made in the same manner as described under *Assay of coenzyme B_{12}*. Finally, the samples were analyzed for the deuterium content by gas chromatography-mass fragmentography, the four ions of masses 465, 466, 467 and 468 being recorded.

The conversion of (*R*)-propane-1,2-diol-1,1-d_2 into propanal by glyceroldehydratase in the presence of non-labeled coenzyme B_{12} in 10^3 times excess over the enzyme, shows that dissociation of the holoenzyme takes place: 29% of coenzyme-5',5'-d_2, about 1% of coenzyme B_{12}-5'-d_1 and 70% of coenzyme B_{12}-d_0 were found after the completion of the enzymic reaction (8°, 3 h). The kinetic studies are in progress.

The technique of gas chromatography-mass fragmentography was also successfully applied to the quantitative analysis of other nucleosides using C-5'-dideuterated nucleosides as internal standard.

EXPERIMENTAL

Low-resolution mass spectra and mass fragmentography of trimethylsilyl derivatives of nucleosides were recorded using a V.G. Micromass Mo. 16 F mass spectrometer (England) coupled with a gas chromatograph and V.G. Data System Mo. 2000. Ionizing energy was 20 eV and ion source and carrier gas separator were set at 230°. Helium was used as a carrier gas at a flow-rate of 35 ml/min. Samples, dissolved in BSTFA, were introduced into the gas chromatograph and separated by using 2.5% SE-30 stationary phase settled on Chromosorb W AW HMDS, 80-100 mesh (100 cm x 2 mm I.D., glass column). The gas chromatographic separations were performed isothermally at 225°.

Coenzyme B_{12}-5',5'-d_2 was synthesized from cyanocobalamin and 2',3'-O-isopropylidene-5'-O-toluene-*p*-sulfonyl-adenosine-5',5'-d_2 ref. 13.

2',3'-O-Isopropylidene-adenosine-5',5'-d_2 was prepared by reduction of 9-(2',3'-O-isopropylidene-β-D-riburonic acid methyl ester)-adenine with $LiAlD_4$ in tetrahydrofuran. Non-labeled coenzyme B_{12} was obtained from Calbiochem (Los Angeles, Calif., U.S.A.). Adenine nucleosides were prepared by photolysis of non-labeled and labeled coenzyme B_{12} and isolated by gel chromatography[4] or thin-layer chromatography (TLC) on silica gel plates (Merck Kieselgel 60 P_{254}). As a solvent for TLC the following mixture was used: chloroform-methanol-30% aqueous ammonia (95:30:1, v/v/v). The R_F values were: hydroxocobalamin and coenzyme B_{12} 0.0, adenosine 0.21, 8,5'-cyclic-adenosine 0.25, 2',3'-O-isopropylidene-adenosine 0.61 and 9-(2',3'-O-isopropylidene-β-D-riburonic acid methyl ester)-adenine 0.63. The glyceroldehydratase was isolated from Aerobacter aerogenes PZH 572. (*R*)-Propane-1,2-diol-1,1-d_2 was prepared by reduction of the methyl ester of (*R*)-lactic acid with $LiAlD_4$ in diethyl ether[14].

ACKNOWLEDGEMENTS

The authors are indebted to Mrs. Maria J. Zagalak for assistance in preparing labeled compounds and Miss Petra Rettig and Miss Susanne Fehlmann for the preparation of the glyceroldehydratase.

REFERENCES

1 R.H. Abeles, C. Myers and T.A. Smith, *Anal. Biochem.*, 15 (1966) 192.
2 M.K. Turner and L. Mervyn, in H.R.V. Arnstein and R.J. Wrighton (Editors), *The Cobalamins, a Glaxo Symposium*, Churchill Livingstone, Edinburgh, 1971, p. 35.
3 H.P.C. Hogenkamp, *J. Biol. Chem.*, 238 (1963) 477.
4 P.Y. Law and J.M. Wood, *Biochim. Biophys. Acta*, 331 (1974) 451.
5 H.P.C. Hogenkamp, J.N. Ladd and H.A. Barker, *J. Biol. Chem.*, 237 (1962) 1950.
6 A.W. Johnson and N. Shaw, *J. Chem. Soc.*, (1962) 4608.
7 P.A. Frey, M.K. Essenberg and R.H. Abeles, *J. Biol. Chem.*, 242 (1967) 5369.
8 J.A. McCloskey, A.M. Lawson, K. Tsuboyana, P.M. Krueger and R.N. Stillwell, *J. Amer. Chem. Soc.*, 90 (1968) 4182.
9 J.J. Dolhun and J.L. Wiebers, *Org. Mass Spectrom.*, 3 (1970) 669.
10 W.A. Koenig, L.C. Smith, P.F. Crain and J.A. McCloskey, *Biochemistry*, 10 (1971) 3968.
11 H. Vogelmann and F. Wagner, *J. Chromatogr.*, 76 (1973) 359.
12 M.K. Essenberg, P.A. Frey and R.H. Abeles, *J. Amer. Chem. Soc.*, 93 (1971) 1242.
13 B. Zagalak and J. Pawelkiewicz, *Acta Biochim. Polon.*, 11 (1964) 49.
14 V. Bonetti, *Dissertation Thesis No. 5366*, Swiss Federal Institute of Technology, Zurich, 1974, p. 72.

MASS FRAGMENTOGRAPHY OF VITAMIN D_3 (CHOLECALCIFEROL) IN PLASMA SAMPLES

A.P. DE LEENHEER and A.A.M. CRUYL

Laboratoria voor Medische Biochemie en voor Klinische Analyse, Faculteit van de Farmaceutische Wetenschappen, RUG, 135 De Pintelaan, 9000 - Gent (Belgium)

SUMMARY

Mass fragmentography has proved to be a powerful tool in the field of biochemistry. The possible application for a quantitative mass--fragmentographic determination of physiological levels of vitamin D_3 in human plasma has been explored. In spite of the specificity claimed for the multiple-ion detection system, the assay procedure necessitates several preliminary clean-up manipulations before mass fragmentographic measurement is possible.

Heparinized plasma is submitted to lipid extraction. The extract is purified from the excess of cholesterol by digitonide precipitation followed by Lipidex^{TM}-5000 column chromatography. Vitamin D_3 is then converted into its isotachysteryl$_3$ heptafluorobutyryl derivative, and an aliquot is injected on top of a 1% FFAP column (2.00 m x 2.0 mm I.D.). Mass-fragmentographic analysis is performed by monitoring simultaneously the molecular ions of isotachysteryl$_3$ heptafluorobutyrate (*m/e* 580, $M^{+\cdot}$) and of dihydrotachysteryl$_2$ heptafluorobutyrate (*m/e* 594, $M^{+\cdot}$); the latter is used as an internal standard.

INTRODUCTION

Several methods have been proposed for the chemical determination of different substances of the vitamin D group. In spite of extensive research there is still need for an analytical method for quantitation of vitamin D in human plasma.

With the recent availability of highly sensitive detection systems, such as electron capture detection, multiple-ion detection and the competitive protein-binding technique, a routine quantitation of vitamin D_3 in human plasma could be developed reaching the low sensitivity levels required.

Wilson *et al.*[1], Sklan *et al.*[2] and Nair *et al.*[3] have described a gas-chromatographic assay with electron capture detection of the heptafluorobutyryl or trifluoroacetyl derivatives of vitamin D_3. Only Belsey *et al.*[4] used the competitive protein-binding technique for the measurement of vitamin D_3 and its liver metabolite 25-hydroxycholecalciferol in plasma. Normal plasma levels of vitamin D_3 with this method ranged from 24-40 ng/ml.

We propose a procedure involving mass-fragmentographic determination of vitamin D_3 in plasma.

MATERIALS AND METHODS

Blood sampling

Blood samples were withdrawn into heparinized tubes. After centrifugation of the blood, 40 mg of ascorbic acid were added to each millilitre of plasma as antioxidant. The plasma samples were stored immediately at -20° until analysed.

Reagents

All chemicals and solvents were of analytical grade and were used without further purification. As reference substances, vitamin D_3 (cholecalciferol) was purchased from Koch-Light (Colnbrook, Great Britain) and dihydrotachysterol$_2$ (DHT_2) from Sigma (St. Louis, Mo., U.S.A.).

Heptafluorobutyric anhydride was obtained from Pierce (Rockford, Ill., U.S.A.). Lipidex^{TM}-5000 came from Packard Instrument (Downers Grove, Ill., U.S.A.) and radioactive [4-^{14}C]vitamin D_3 (spec. act. 23.5 mCi/mmole) from Philips-Duphar (Amsterdam, The Netherlands). 2,6-Di-*tert*.-butyl-4-methyl phenol (BHT, butylated hydroxytoluene) was obtained from Aldrich-Europe (Beerse, Belgium).

Mass-fragmentography instrumentation

Mass-fragmentographic analysis was performed on an LKB 9000S combined gas chromatograph and mass spectrometer (GC/MS) electron-impact mode equipped with a multiple-ion detection (MID) device. The analyses were carried out on a silanized glass column (2.00 m x 2.0 mm I.D.) packed with 1% FFAP Gas-Chrom Q (100-120 mesh). The temperature of the injection port and the oven were kept at 230-250° and 215°, respectively. The flow--rate of carrier gas (helium) was 30 ml/min. The temperature of the double Becker-Ryhage jet separator was maintained at 275° and the ion source temperature at 270°. The ionization potential was set to 20 eV and the trap current to 60 μA. The MID served as a quantitative specific ion detector for the gas chromatograph. The molecular ions of isotachysteryl$_3$ heptafluorobutyrate (ISOT$_3$-HFB) and dihydrotachysteryl$_2$ heptafluorobutyrate (DHT$_2$-HFB) were selected for quantitative monitoring. The first channel of the MID unit was focused on the ion at *m/e* 580 (ISOT$_3$-HFB, $M^{+\cdot}$), whereas for the second ion at *m/e* 594 (DHT$_2$-HFB, $M^{+\cdot}$), an additional accelerating voltage of 417.25 V was applied.

The gain of both channels was adjusted to give a relative intensity 1:1. The filter setting was 0.6 Hz and the measuring time was 20 msec.

Lipidex column chromatography

Lipidex^{TM}-5000 column chromatography was carried out in silanized glass columns (25 cm x 4 mm I.D.) equipped with a solvent reservoir and a tap. Lipidex^{TM}-5000 was equilibrated for 30 min in methanol containing 0.01% BHT which was also used as the eluent; the column was packed under gravity flow. After use the column was washed extensively with methanol. The same column can be re-used indefinitely provided it is not allowed to dry out.

Procedure

(a) *Lipid extraction*. Plasma samples are extracted according to a modified Bligh-Dyer procedure (*cf*. ref. 5). Radioactive [4-^{14}C]-vitamin D$_3$ (*ca*. 700 dpm) is added to 5 ml of plasma (to assess each time the efficiency of the extraction and clean-up) and left to equilibrate for 30 min at room temperature. After successive additions of 5.0 ml of water, 12.5 ml of chloroform and 25.0 ml of methanol containing 0.01% BHT as antioxidant, the homogeneous solution is mixed and left to digest for 30 min at room temperature. Another 10 ml of water and 12.5 ml of chloroform are added, and without further mixing the samples are centrifuged for 15 min at 2000 rpm. The centrifuged mixture consists of an organic chloroform layer at the bottom, an aqueous methanol phase at the top and a fine precipitate of denatured proteins at the interphase. The chloroform layer is collected and the aqueous phase re-extracted with 25 ml of chloroform.

The combined chloroform extracts are evaporated to dryness under vacuum at a temperature below 40°.

(b) *Digitonide precipitation.* The residue is dissolved in 20 ml of 96% ethanol containing 0.01% BHT and is slightly warmed until clear. Then 8.6 ml of water and 10 ml of 2% digitonin in 72% (by volume) ethanol are added and the mixture is allowed to stand overnight at 4°. The precipitate formed is centrifuged off for 15 min at 3000 rpm. The supernatant is extracted twice with 25 ml of light petroleum (b.r. 40-60°). The formation of an inseparable emulsion may be avoided by shaking the mixture very gently. The combined extracts are washed with 20 ml of 72% ethanol and evaporated to dryness under vacuum at a temperature below 40°.

(c) *Lipidex column chromatography.* The purified extract is dissolved in 0.2 ml of methanol containing 0.01% BHT and applied to the column. Elution is carried out at atmospheric pressure and room temperature with methanol as eluant at a flow-rate of about 4 ml/h. The fraction containing vitamin D_3 is collected. About 40 ng of DHT_2 is added to the eluate as an internal standard. The eluate is evaporated to dryness under vacuum at a temperature below 40°.

(d) *Derivatization procedure.* The evaporated eluate is dissolved in 200 µl of benzotrifluoride and derivatized with 5 µl HFBA for 30 min at room temperature. The solvent is removed with the aid of a stream of purified nitrogen. The residue is dissolved in 50 µl of chloroform, and an aliquot of about 1 µl is injected into the GC/MS.

RESULTS AND DISCUSSION

The modified Bligh-Dyer extraction for lipids gave the best results among different lipid extraction procedures investigated. Forming a monophase system before extraction assures an optimal contact between the plasma and methanol as protein denaturing solvent and also with chloroform as lipid extractant. Subsequent complete separation of the different layers eliminates the water-soluble constituents from the plasma.

In spite of the specificity claimed for the multiple-ion detection, an extensive purification of the plasma was found to be necessary. Cholesterol and an unknown high molecular lipid constituent of the plasma are the major interfering components. The concentration ratio of cholesterol to vitamin D_3 in normal plasma amounts to about 10^5:1, and the two compounds have only 2 a.m.u. difference. The mass spectra of ISOT$_3$--HFB (vitamin D_3 derivative) and cholesteryl heptafluorobutyrate (CHOL-HFB) are shown in Figs. 1 and 2. The M-2 peak of CHOL-HFB interferes with the monitored ion (*m/e* 580) in the vitamin D_3 assay. For a successful determination the amount of cholesterol in the extract must be sufficiently reduced, because, the very similar chemical reactivity of cholesterol and vitamin D_3 hampers the separation of the two compounds. In a first step a crude elimination of cholesterol is achieved by forming an insoluble equimolar digitonide complex[6], typical for 3β-hydroxysteroids. (The insertion of a methyl group at C-10 in epi-configuration prevents the formation of an insoluble complex for vitamin D_3.) The extract is further purified on Lipidex^{TM}-5000. This 50% alkylated hydroxyalkoxypropyl Sephadex LH-20 derivative is a very appropriate gel for separating non--polar lipids. However, vitamin D_3, being very sensitive to oxidation, requires the addition of BHT to the eluant. Indeed, Lipidex^{TM}-5000 has a tendency to form peroxides if it remains in contact with methanol for long and the omission of an antioxidant leads to extensive degradation of vitamin D_3 on the columns. Apart from this inconvenience, Lipidex^{TM}-5000 is inert, very stable and gave very reproducible results over a one-year analysis period. Lipidex column chromatography gives the best separation of vitamin D_3 and cholesterol compared with other tested adsorption and partition chromatographic purifications. As an example the chromatographic separation of [4-^{14}C]vitamin D_3 and cholesterol is shown in Fig. 3.

Finally, complete separation from the remaining cholesterol is obtained on FFAP as the gas-chromatographic liquid stationary phase. Fig. 4 shows the MID recordings of different concentrations of pure standard samples after Lipidex column chromatography (30, 60 and 90 ng of vitamin D_3). Figs. 5 and 6 give the mass-fragmentographic recordings of extracts of plasma, one spiked with 24 ng vitamin D_3 per millilitre and another without added vitamin D_3, respectively.

As seen in Fig. 6, a peak occurs with a retention time slightly lower than that of vitamin D_3 itself. This unknown peak becomes more significant when higher sensitivity is required to determine lower plasma levels. Quantitation based upon peak height measurement proved to be accurate; thus a better separation is not required. The over-all recovery of radioactively labeled vitamin D_3 - when added to plasma and taken through the entire procedure, *i.e.* solvent extraction, digitonide precipitation, Lipidex chromatography and derivatization - was 76.3% (C.V., 6.9%; n = 30).

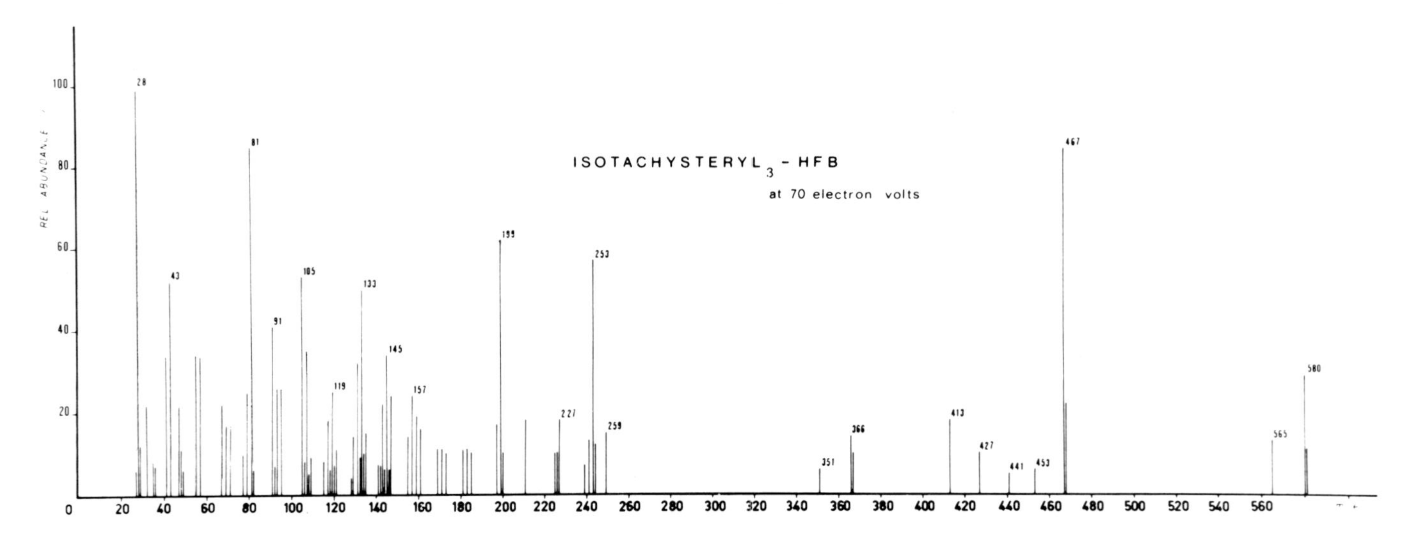

Fig. 1. Mass Spectrum of isotachysteryl$_3$ heptafluorobutyrate (ISOT$_3$-HFB, vitamin D_3 derivative).

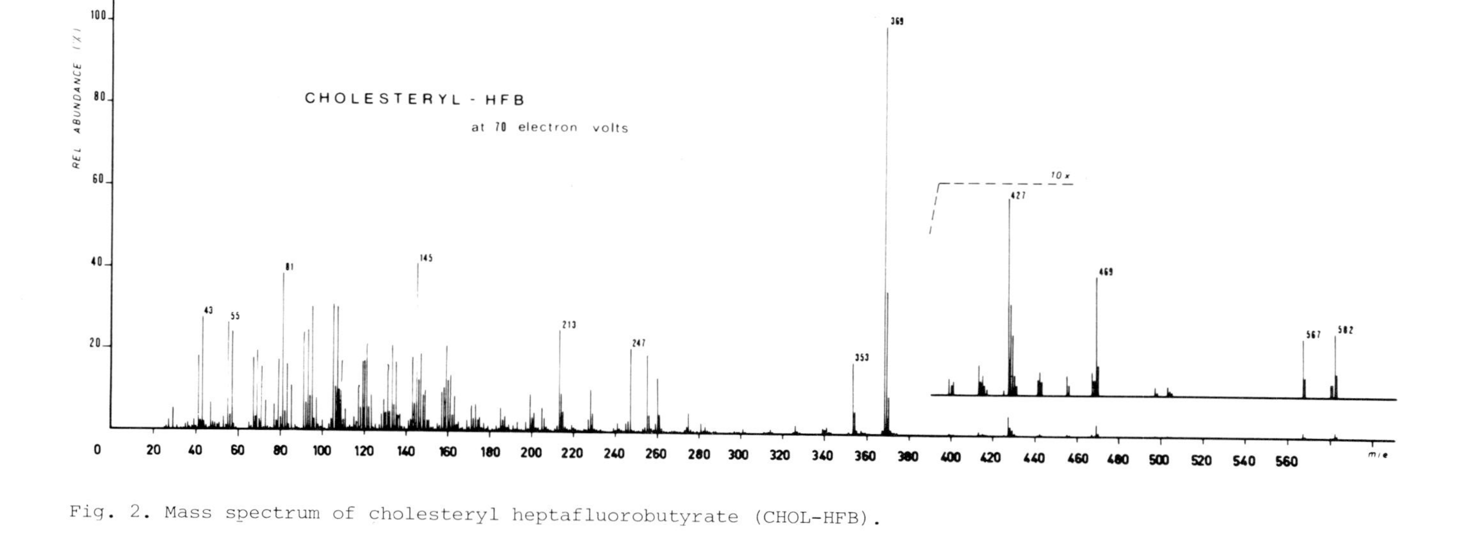

Fig. 2. Mass spectrum of cholesteryl heptafluorobutyrate (CHOL-HFB).

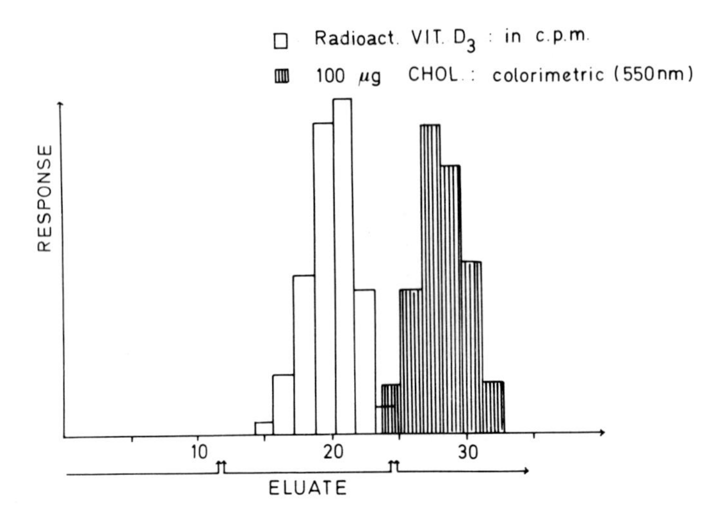

Fig. 3. Lipidex column chromatography of a mixture of [4-C^{14}]vitamin D_3 and cholesterol.

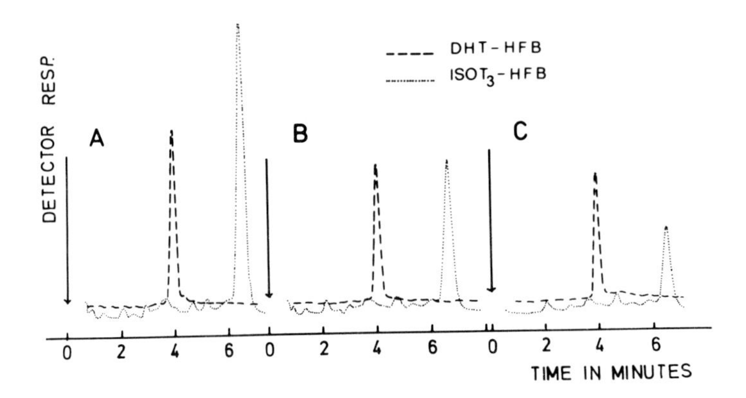

Fig. 4. MID recordings of the ions at *m/e* 580 and 594 of $ISOT_3$-HFB and DHT_2-HFB, respectively, after Lipidex column chromatography of different amounts of vitamin D_3. A, 90 ng; B, 60 ng; and C, 30 ng.

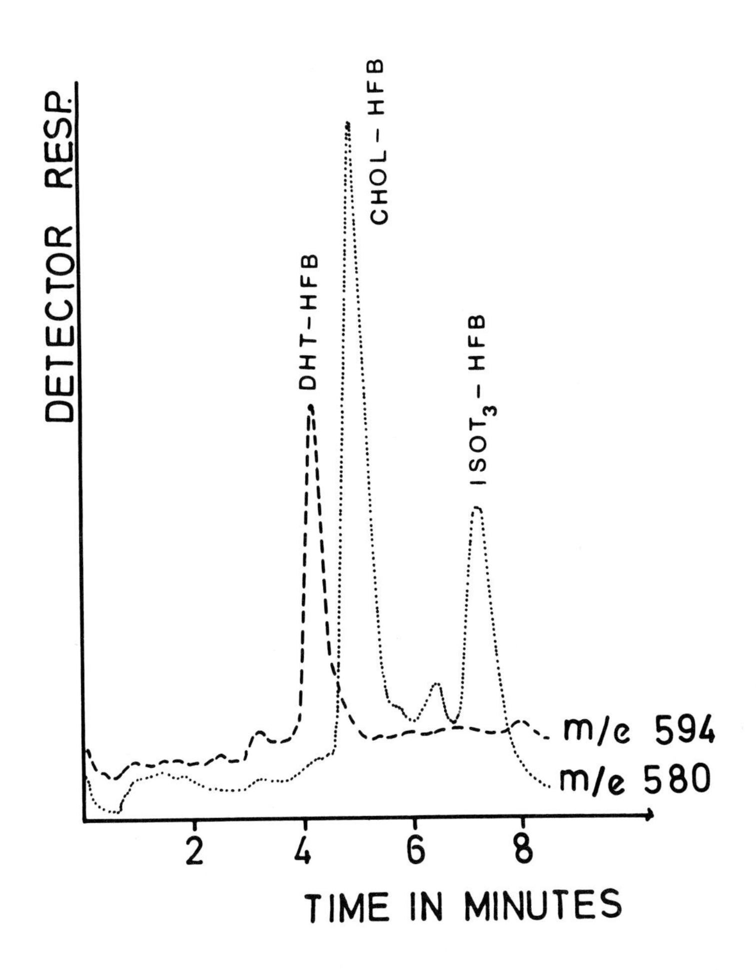

Fig. 5. MID recording of a purified heptafluorobutyryl derivatized extract of plasma, spiked with vitamin D_3 (24 ng/ml plasma).

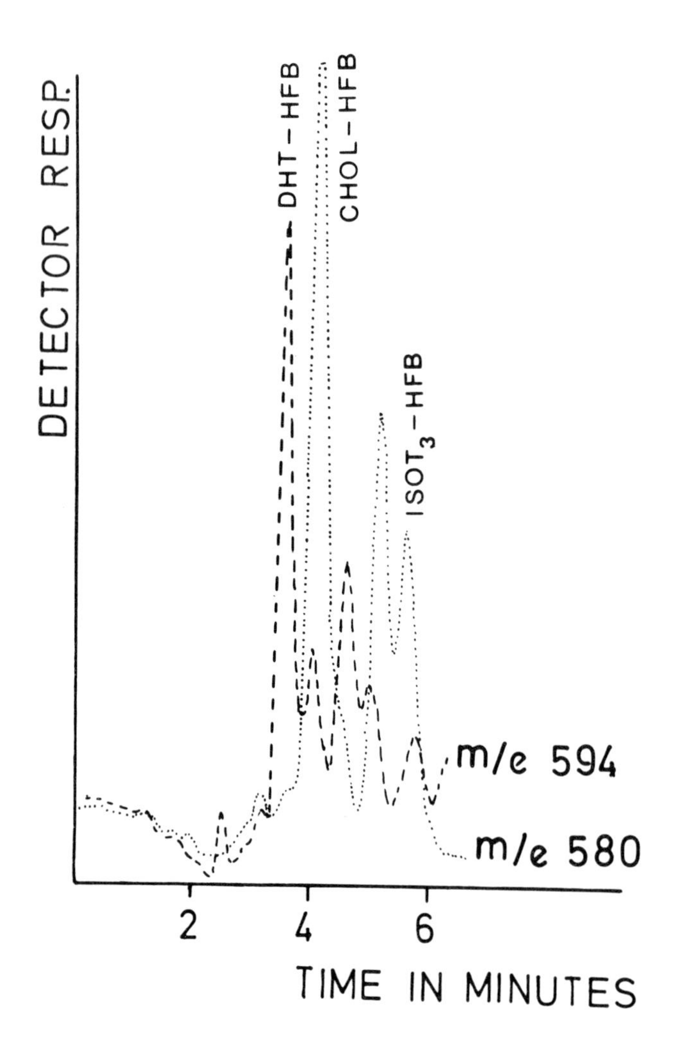

Fig. 6. MID recording of a purified heptafluorobutyryl derivatized extract of plasma.

CONCLUSION

Major problems in the determination of vitamin D_3 in plasma have to be overcome. The high instability of the compound itself dictates mild and cautious working conditions (avoidance of daylight, use of an antioxidant, etc.). The thermal decomposition of vitamin D_3 into its pyro- and isopyrocalciferol isomers upon gas-chromatographic injection is avoided by conversion into its isotachysteryl$_3$ heptafluorobutyryl derivative (ISOT$_3$--HFB). The extremely high amounts of cholesterol occurring in plasma are substantially reduced before mass-fragmentographic analysis. The presence in the purified plasma extract of another unidentified constituent with similar gas-chromatographic properties to those of ISOT$_3$-HFB limits the sensitivity of the method. A more pronounced clean-up of the Lipidex eluate is necessary for the analysis of subnormal plasma levels of vitamin D_3.

ACKNOWLEDGEMENTS

This work was supported in part by the National Research Foundation (NFWO) through a bursary (aspirant) to one of us and the Foundation for Medical Scientific Research (FGWO) by their grants 3.0001.76, 20 210 and 20 007.

REFERENCES

1 P.W. Wilson, D.E.M. Lawson and E. Kodicek, *J. Chromatogr.*, 39 (1969) 75.
2 D. Sklan, P. Budowski and M. Katz, *Anal. Biochem.*, 56 (1973) 606.
3 P.P. Nair and S. de Leon, *Arch. Biochem. Biophys.*, 128 (1968) 663.
4 R.F. Belsey, H.F. DeLuca and S.T. Potts, *J. Clin. Endocrinol. Metab.*, 33 (1974) 554.
5 M.A. Preece, J.L.H. O'Riordan, D.E.M. Lawson and E. Kodicek, *Clin. Chim. Acta*, 54 (1974) 235.
6 E.J. de Vries, F.J. Mulder and K.J. Keuning, *J. Vitaminol.*, 15 (1969) 189.

QUALITATIVE AND QUANTITATIVE ANALYSIS OF DOPAMINE BY ELECTRON IMPACT AND FIELD DESORPTION MASS SPECTROMETRY

W.D. LEHMANN, H.D. BECKEY and H.-R. SCHULTEN

Institute of Physical Chemistry, University of Bonn, Wegelerstr. 12 5300 Bonn (G.F.R.)

SUMMARY

The derivatives of dopamine, 3-O-methyldopamine, and 4-O--methyldopamine produced by reaction with 5-di-*n*-butylaminonaphthalene-1--sulphonyl chloride were investigated by electron impact and field desorption mass spectrometry. This derivatization technique made possible the detection and quantitation of dopamine and its monomethyl ethers in human urine. The quantitation was achieved by two independent methods. (1) A twin direct introduction system for calibration in the electron impact mode and (2) stable isotope dilution using electron impact and field desorption mass spectrometry. Further, the utility of field desorption mass spectrometry for quantitative determination of underivatized dopamine by stable isotope dilution is demonstrated. The estimation of dopamine at concentrations of 0.1-10 μg/ml could be performed with an error between 5 and 10%.

INTRODUCTION

Although many biologically active amines (biogenic amines) are known in recent years analytical interest has been focused on a relatively small number of these compounds, in particular, on those amines that act as inhibitors or transmitters of signals in the nervous system. These include small aliphatic amines (quaternary ammonium bases) such as acetylcholine and phenolic or catecholic amines such as tyramine and dopamine. When mass spectrometry (MS) is included in the analysis of biogenic amines, in general derivatives are produced and subsequently estimated by a coupled gas chromatograph and mass spectrometer unit[1,2].

Ions with *m/e* values considerably below the molecular weight must be used for identification, since these compounds mostly undergo strong fragmentation under electron impact (EI). In this paper we introduce a derivatization technique for dopamine and its methyl ethers that yields derivatives amenable to qualitative analysis by EI and field desorption (FD) MS[3]. Further, we give one of the first quantitative results obtained in the FD mode with isotopically labelled compounds as internal standards.

EXPERIMENTAL

Dopamine (DA) and 4-O-methyldopamine (4-MDA) as hydrochlorides were purchased from Aldrich (Milwaukee, Wisc., U.S.A.). 3-O-Methyldopamine (3-MDA) hydrochloride was obtained from Regis (Morton Grove, Ill., U.S.A.). Dopamine-(α-d_2, β-d_2) hydrochloride was obtained from Merck, Sharp and Dohme, Canada. The compounds were of analytical grade. 5-Di-*n*--butylaminonaphthalene-1-sulphonyl chloride (BANS-Cl)[4,5] was kindly supplied by N. Seiler, Max-Planck-Institut für Hirnforschung, Frankfurt, G.F.R. The reaction of the amines with BANS-Cl was carried out according to a procedure described by Seiler[6] for 5-dimethylaminonaphthalene-1-sulphonyl chloride (DANS-Cl). Thin-layer chromatography (TLC) was carried out on glass plates (20 cm x 20 cm) coated with 300 μm silica gel G layers (Merck, Darmstadt, G.F.R.) using ascending chromatography. The thin-layer plates were developed with cyclohexane-ethyl acetate (4:1) as eluent. The fluorescent spots (excitation at 364 nm) were scraped off and extracted with 500 μl ethyl acetate. After appropriate concentration under a stream of nitrogen, the extract was transferred into the microcrucible of the direct introduction system of the EI mass spectrometer. Alternatively, for FD analysis the extract was applied to a high temperature activated wire emitter by the syringe technique[3].

The EI spectra and the FD spectra using electric detection were recorded on a modified CH 4 mass spectrometer. The EI spectra were run under standard conditions (70 eV electron energy, 20 μA emission, 150° ion source temperature). The ion source of the mass spectrometer was equipped with two identical direct introduction systems facing each other (twin system). Experimental details are given elsewhere[7]. The FD spectra detected photographically were run on a double-focusing CEC21-110B mass spectrometer. Field anodes used were 10-μm tungsten wires activated at high temperature[8]. The average length of the micro-needles was about 30 μm. The ions were recorded on vacuum-evaporated AgBr photoplates (Ionomet, Waban, Mass., U.S.A.). All spectra were produced by emission-controlled FD MS[9].

RESULTS AND DISCUSSION

Qualitative analysis of bansyl derivatives of dopamine and its monomethyl ethers

It is well known that DANS-Cl reacts very easily with molecules having primary or secondary amino groups or phenolic hydroxyl functions. BANS-Cl is a modified DANS-Cl containing *n*-butyl groups in place of the methyl substituents on the amino group. If the derivatives formed by these reagents are identified by EI MS the use of BANS-Cl offers an essential advantage: the base peak in the mass spectra of the DANS derivatives is a non-specific fragment ion at *m/e* 170[10,11], whereas in the mass spectra of the BANS derivatives the base peak is formed by cleavage of C_3H_7 from the molecular ion[4]. In addition the relative intensities of the molecular ions for the DANS derivatives are in the order of a few percent only (especially in multiply substituted derivatives). The BANS derivatives, however, show molecular ions of 30-40% relative intensity on average.

The result of the reaction of DA with BANS-Cl was a mixture of four different derivatives: one monobansyl-DA, two constitutionally isomeric bisbansyl-DA derivatives, and one trisbansyl-DA. Owing to their differences in polarity the reaction products are easily separated by TLC and are clearly located by their strong UV fluorescence. Under the conditions selected the tris-BANS-DA represented the main product. As shown in Fig. 1, the EI spectra of all four derivatives yielded the $(M-43)^{+\cdot}$ ion as the base peak. The bisbansyl and trisbansyl derivatives of DA were isolated from human urine samples which were spiked with DA at a level of 10 µg/ml. This value is in the order of magnitude of the concentration of unconjugated DA which is excreted in the urine of patients with Parkinson's disease undergoing L-Dopa therapy[12].

The reaction of 3-MDA and 4-MDA with BANS-Cl produced a mixture of two different derivatives: one monobansyl and one bisbansyl derivative. The bisbansyl derivatives of 3-MDA and 4-MDA were separated on TLC nearly completely (R_F = 0.31 and 0.34, respectively), whereas their monobansyl derivatives exhibited identical R_F values (0.17). As in the derivatization of dopamine, the reaction with BANS-Cl was performed with a large excess of this reagent and yielded the bisbansyl derivatives as the main products. The EI spectra of the four derivatives displayed the $(M-43)^{+\cdot}$ ion as the base peak and the molecular ions with relative abundances between 30 and 47%. With the use of FD MS only, the molecular ions and no fragment ions were observed. The bisbansyl derivatives were isolated from human urine samples that contained 3-MDA and 4-MDA in concentrations of 10 µg/ml.

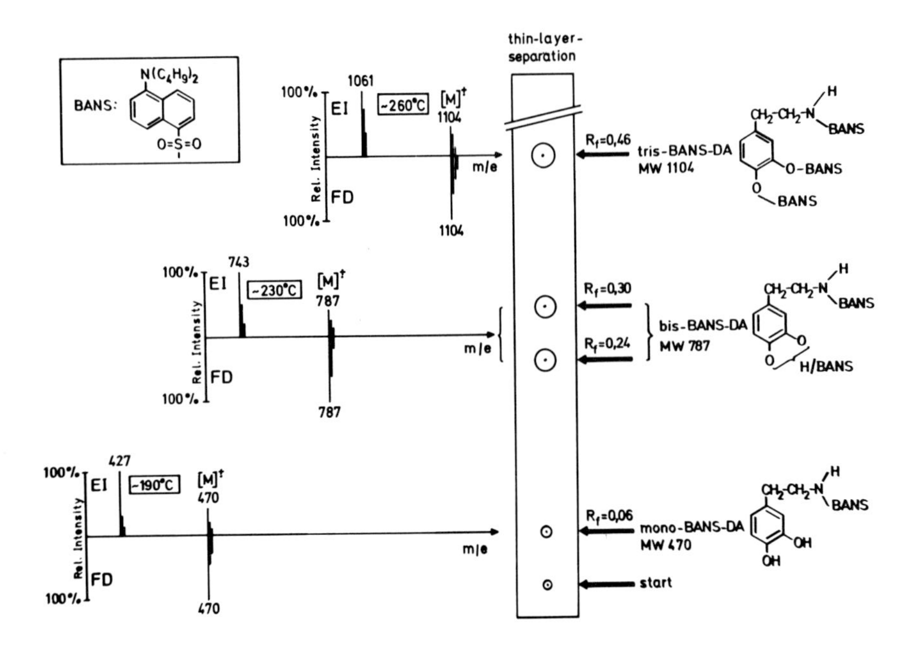

Fig. 1. Thin-layer separation of the four derivatives that are formed by reaction of BANS-Cl with DA and partial EI and FD mass spectra of the separated reaction products. The mass range from *m/e* 300-1200 was recorded electrically with EI MS. The FD spectra were detected on the photoplate in the mass range from *m/e* 40-1400 (ref. 7).

Quantitation of BANS derivatives

For quantitative determination it is advantageous to use the trisbansyl derivative of DA and the bisbansyl derivatives of 3-MDA and 4-MDA since: (1) these derivatives have the lowest polarity and are therefore easily separated from the multiplicity of polar substances in urine. Their extraction from the reaction mixture can be performed with *n*-heptane. (2) Further, these derivatives have the highest molecular weight, thus diminishing background problems and (3) the highest sensitivity is observed for these derivatives because they are main products in the reaction mixtures. For the estimation of the bansyl derivatives two different, independent MS methods have been used: a twin direct introduction system with EI MS and stable isotope dilution in connection with EI and FD MS.

The EI mass spectrometer was equipped with identical direct introduction probes facing each other. The system was calibrated by determining the relative sensitivity coefficient (sensitivity direct inlet(1)/sensitivity direct inlet(2)) in 20 measurements to be 1.17 ± 0.10. For the estimation of DA, 3-MDA and 4-MDA the samples were split into two halves, and to one of them a known amount of the compound to be assayed was added as external standard. The authentic sample (1) and the one to which the standard was added (2) were subjected to the same work-up procedure. The isolated BANS derivatives of samples (1) and (2) were transferred into the micro-ovens of direct inlets(1) and (2) and evaporated consecutively. During evaporation the mass spectrometer is set on a fixed *m/e* value (single ion monitoring), *e.g.* the $(M-43)^{+\cdot}$ ion of a BANS derivative. The evaporation profiles of bisbansyl 3-MDA, selected ion *m/e* 758 (corresponding to $(M-43)^{+\cdot}$), for a quantitation are shown in Fig. 2. Since the relative sensitivity coefficient of direct inlet (1) and (2) is known, the amount of 3-MDA present in the sample can be calculated in the following way.

$$\frac{\text{peak area (1)}}{\text{peak area (2)}} = \frac{\text{sample amount (1)}}{\text{sample amount (2)}} \times 1.17$$

This method of using an external standard yields quantitative results for concentrations between 0.1 and 10 µg/ml urine with an average error of ± 10%.

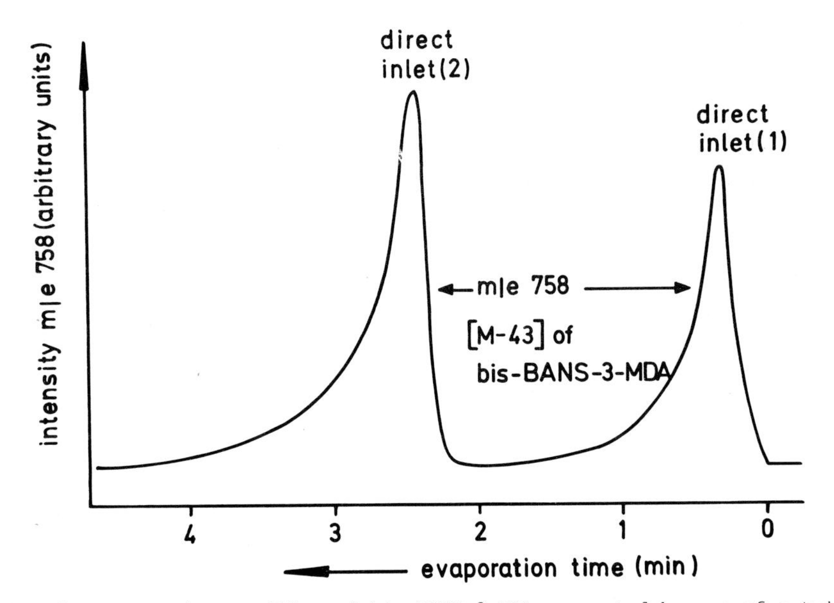

Fig. 2. Evaporation profiles of bis-BANS-3-MDA generated by use of a twin direct introduction system and EI MS. The probes were heated from ambient temperature to about 500° within two min (ref. 7).

The method of stable isotope dilution in connection with mass fragmentography has been introduced as the method of choice for the assay of drugs[13,14]. For the estimation of DA in urine we have used the trisbansyl derivatives of DA-d_0 and DA-d_4. With EI MS at low resolution the estimation of the trisbansyl derivative in the range as described above could be performed with an error of about $\pm$ 5%. Additionally, FD MS was used for measurement. Photographic detection of the molecular ion group of a mixture of trisbansyl-DA-d_0 and -d_4 in the molar ratio 1:0.95 gave the spectrum displayed in Fig. 3. The theoretical molar ratio was reproduced within an error of 5% (the sample consumption was about 200 ng each). The accuracy is consistent with a determination of a mixture of cyclophosphamide-d_0 and -d_6 with FD MS and photoplate detection reported previously[15].

Quantitation of underivatized catecholamines

Since the FD mass spectra of a wide variety of organic compounds exhibit high molecular ion intensities besides weak or no fragment ions, FD MS is well suited in principle for the detection of compounds without

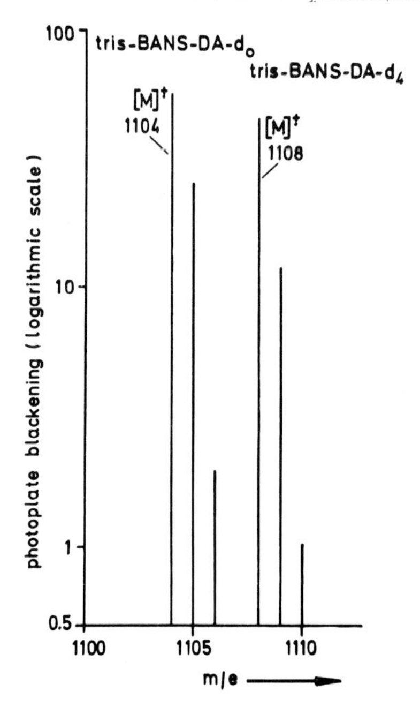

Fig. 3. FD mass spectrum of trisbansyl-DA and trisbansyl-DA-d_4 (molar ratio 1:0.95). Total sample amount (d_0 + d_4) about 400 ng. Emission controlled desorption, threshold 1 x 10^{-8} A, exposure time of the photoplate 19 min.

derivatization[3]. For example, the FD spectra of many biogenic amines yield the molecular ion as base peak[16,17], whereas in EI MS the molecular ion intensity is often in the order of a few percent. Therefore the potential of FD for the measurement of underivatized DA by stable isotope dilution using photographic and electric detection was investigated. Two basic requirements for the effective use of FD MS in quantitative investigations are fulfilled. (1) The possibility of producing emitters of uniform emission qualities and (2) the adjustment of reproducible conditions during the run of an FD spectrum, *e.g.* by emission-controlled desorption. Fig. 4 shows the profiles of the emitter heating current and the total emission during three different runs of an FD spectrum using three different emitters.

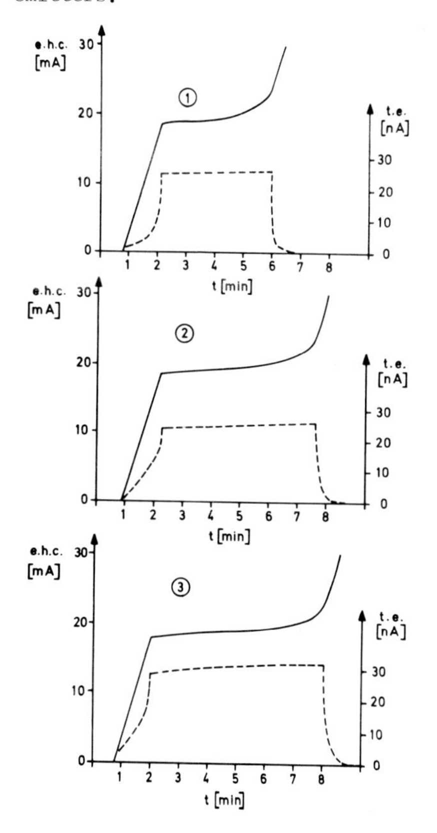

Fig. 4. Emitter heating current (e.h.c.) (——) and total emission (t.e.) (---) during three different FD runs of a mixture of DA-d_0/-d_4 with three different emitters and emission-controlled desorption. Total emission threshold: runs 1 and 2, 26 nA; run 3, 30 nA.

The runs of the FD spectra shown in Fig. 4 were detected photographically, and the results of the quantitative determinations are given in Table I. The theoretical molar ratio of DA-d_0:DA-d_4 was reproduced with an error of 5%. The measurement of DA with repetitive scan over the molecular ion group of a mixture of DA-d_0 and -d_4 with electric detection during a single desorption process yielded a somewhat smaller deviation of 3% from the true value (see Table II).

Table I

Quantitation of dopamine-d_0/-d_4 by FD MS with photographic detection

The profiles of the emitter heating current and the total emission corresponding to the three spectra are shown in Fig. 4.

m/e	rel.int.1	rel.int.2	rel.int.3
153	100	100	100
154	36.5	56.2	41
157	46.6	47.7	48.8
158	22.2	27.9	19.3

molar ratio d_0/d_4 (exp.), 1:0.490; (theor.), 1:0.465; error, +5%.

Table II

Quantitation by FD MS using electric detection

Emitter heating current, 17 mA. Electric detection averaged over 22 scans.

molar ratio d_0/d_4 (exp.)	1:0.970 $\pm$ 0.05	error: -3%
molar ratio d_0/d_4 (theor.)	1:1	

CONCLUSION

The reaction of DA, 3-MDA and 4-MDA with BANS-Cl yields defined derivatives of low polarity, good chemical stability and high molecular weight. From human urine these can be measured, with an external standard, by EI MS (in connection with a twin direct inlet system) or by the use of an internal standard by EI and FD MS (isotope dilution technique). Isotopically labelled compounds can successfully be used for measurement by FD MS with photographic and electric detection. The results show the potential of this novel technique for physiological and pharmacokinetic studies.

ACKNOWLEDGEMENTS

This work was supported by grants from the Deutsche Forschungsgemeinschaft, Landesamt für Forschung des Landes Nordrhein-Westfalen and Fonds der Deutschen Chemischen Industrie.

REFERENCES

1 S.H. Koslow, F. Cattabeni and E. Costa, *Science*, 176 (1972) 177.
2 E. Änggard and G. Sedvall, *Anal.Chem.*, 41 (1969) 1250.
3 H.D. Beckey and H.-R. Schulten, *Angew.Chem.Int.Ed.Engl.*, 14 (1975) 403.
4 N. Seiler, T. Schmidt-Glenewinkel and H.H. Schneider, *J.Chromatogr.*, 84 (1973) 95.
5 N. Seiler and H.H. Schneider, *Biomed.Mass Spectrom.*, 1 (1974) 381.
6 N. Seiler, *Methods Biochem.Anal.*, 18 (1970).
7 W.D. Lehmann, H.D. Beckey and H.-R. Schulten, *Anal.Chem.*, in press.
8 H.-R. Schulten and H.D. Beckey, *Org.Mass Spectrom.*, 6 (1972) 885.
9 H.-R. Schulten and H.D. Beckey, *Recent Advances in Field Desorption Mass Spectrometry, 23rd Ann.Conf. on Mass Spectrometry and Allied Topics, Houston, Texas, May 25-30, 1975,* Conf.Proceedings B-1.
10 C.R. Creveling, K.Kondo and J.W. Daly, *Clin.Chem.*, 14 (1968) 302.
11 N. Seiler, H. Schneider and K.-D. Sonnenberg, *Z.Anal.Chem.*, 252 (1970) 127.
12 R.L. Bronaugh, R.J. McMurtry, M.M. Hoehn and C.O. Rutledge, *Biochem. Pharmacol.*, 24 (1975) 1317.
13 C.C. Sweeley, W.H. Elliot, I. Fries and R. Ryhage, *Anal.Chem.*, 38 (1966) 1549.
14 C.G. Hammar, B. Holmstedt and R. Ryhage, *Anal.Biochem.*, 25 (1968) 532.
15 H.-R. Schulten, *Cancer Treatment Rep.*, 60 (1976) 501.
16 G.W. Wood, N. Mak and A.M. Hogg, *Anal.Chem.*, 48 (1976) 981.
17 H.-R. Schulten, *Methods Biochem.Anal.*, 24 (1977).

QUANTITATIVE FIELD DESORPTION MASS SPECTROMETRY

I. CYCLOPHOSPHAMIDE

H.-R. SCHULTEN and W.D. LEHMANN

Institute of Physical Chemistry, University of Bonn, Wegelerstr. 12, 5300 Bonn (G.F.R.)

and

M. JARMAN

Chester Beatty Research Institute, Institute of Cancer Research, Royal Cancer Hospital, Fulham Road, London SW3 6JB (Great Britain)

SUMMARY

Cyclophosphamide, a widely used antitumour agent, has been studied by quantitative field desorption mass spectrometry with stable isotope dilution. The accuracy and precision of the data obtained by electric and photographic detection are reported. Analysis in the microgram and submicrogram range revealed deviations between 2 and 10% from the true value. The low-resolution measurements gave coefficients of variation between 2 and 4%.

INTRODUCTION

Molecular weight determination of thermally labile and non-volatile substances by field desorption mass spectrometry (FD MS) have established the method as a powerful tool in mass spectrometric analysis[1-3]. The characteristic features of the method for the qualitative analysis of a wide variety of model compounds have been demonstrated. However, quantitative determinations by FD MS are handicapped because of the following variable factors:

(1) the quality of the field desorption emitter;

(2) the adjustment of reproducible desorption conditions;

(3) the transfer of the sample onto the FD emitter;

(4) the influence of chemical properties of the compound under investigation on the desorption process.

RESULTS AND DISCUSSION

The production of emitters of similar field desorption properties can be achieved by high-temperature activation in a multiple activation apparatus, control of the shape, distribution and length of the micro--needles under a light microscope, and testing of the ionization effiency of the emitter in the field ionization mode.

Reproducible desorption conditions can be achieved when emission--controlled FD MS is used as described recently[4].

In our experience the syringe technique enables the reliable transfer of defined amounts of sample onto the emitter[5].

The virtually identical chemical properties of compounds and their isotopically labelled analogues allow quantitative FD MS analysis in the isotope-dilution technique.

Concurrent with the improvements of the FD technique a major aim in our research programme has been the utilization of FD MS for analytical problems in cancer research. These studies were concentrated on alkylating antitumour agents such as cyclophosphamide (CP) (I), isophosphamide, trophosphamide and some metabolites of CP[4,6]. Since CP is itself extensively fragmented in the electron-impact ionization mode, and since it gives rise to thermally unstable (*e.g.* 4-hydroxy-cyclophosphamide, carboxyphosphamide (III)) as well as ionic, and hence non-volatile, metabolites (*e.g.* phosphorodiamidic acid (IV)), the FD technique has potential value for the detection and quantitation of the drug and its metabolites in underivatized mixtures.

Fig. 1 shows the structural formulae of compounds I-IV (IV as its cyclohexylammonium salt) and the FD spectrum of an equimolar mixture of these compounds. The molecular ions or their protonated counterparts are displayed for all four compounds with high relative intensity. Thus the capacity of the method for analysing mixtures of compounds which differ widely in polarity is demonstrated. The successful detection of CP and some of its metabolites in this qualitative experiment prompted us to investigate the quantitation of CP as a first step towards the aim of quantitating CP and its metabolites in biological samples.

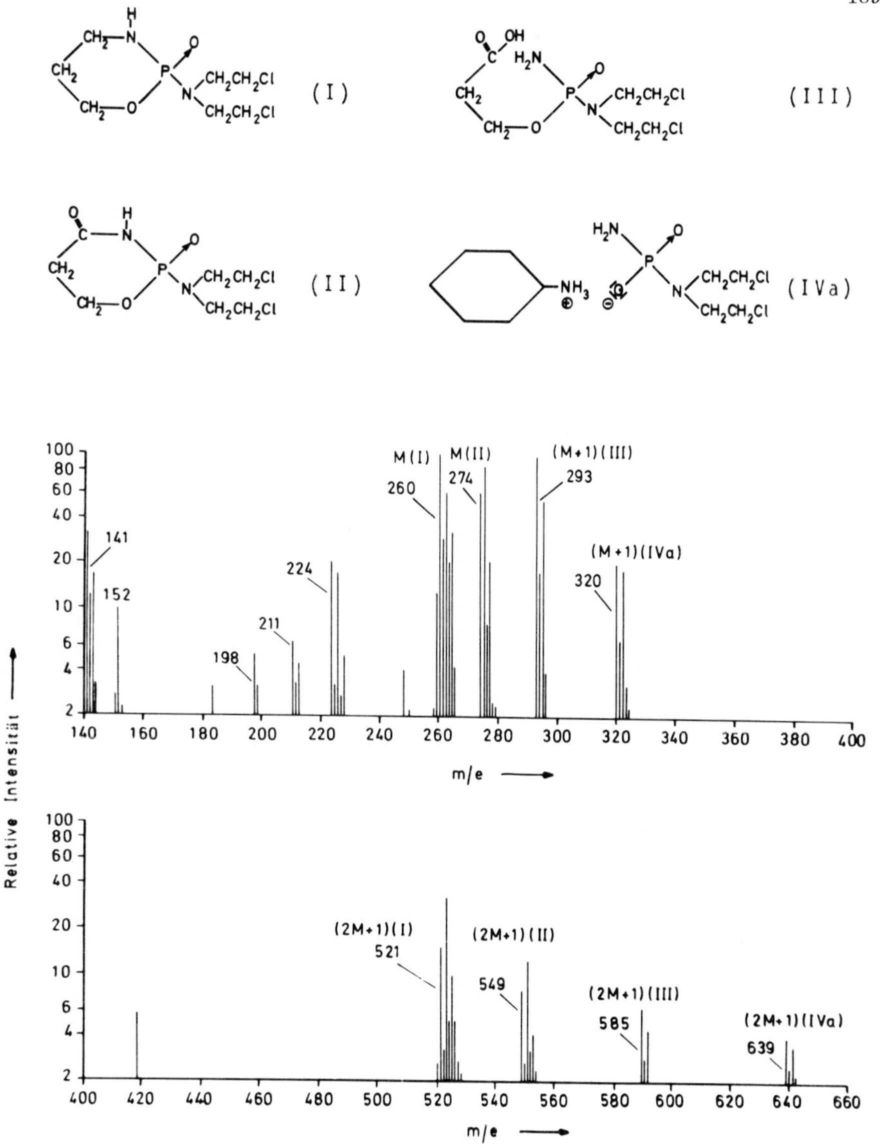

Fig. 1. Field desorption mass spectrum of a mixture of compounds I-IV. A solution/suspension of 50 µg of each compound in 10 µl acetone was used. The amount of sample adsorbed on the emitter surface and consumed for one FD mass spectrum was estimated and calculated to be less than 0.1 µg (ref. 6).

Table I

Quantitation of mixtures of cyclophosphamide-d_0/d_6 with *photographic detection* by FD MS

Blackening of the photoplate; I(exp.), found relative intensity; I(theor.), calculated relative intensity. The isotopic distribution of the standard cyclophosphamide-d_6 was determined by EI MS to be $d_4:d_5:d_6$ = 3:20:100.

m/e	B	I(exp.)(%)	I(theor.)(%)	Error (%)	Mixture
260	0.038	5.5	4.9	+ 12.2	5:100
266	1.171	100	100		
260	0.911	-	-	- 10.8	20:100
261	0.217	17.8	20		
266	-	-	-		
267	1.367	100	100		
260	0.393	100	100	+ 12.6	100:100
266	0.441	114	101.2		
260	1.098	100	100	- 6.1	500:100
266	0.248	19	20.2		

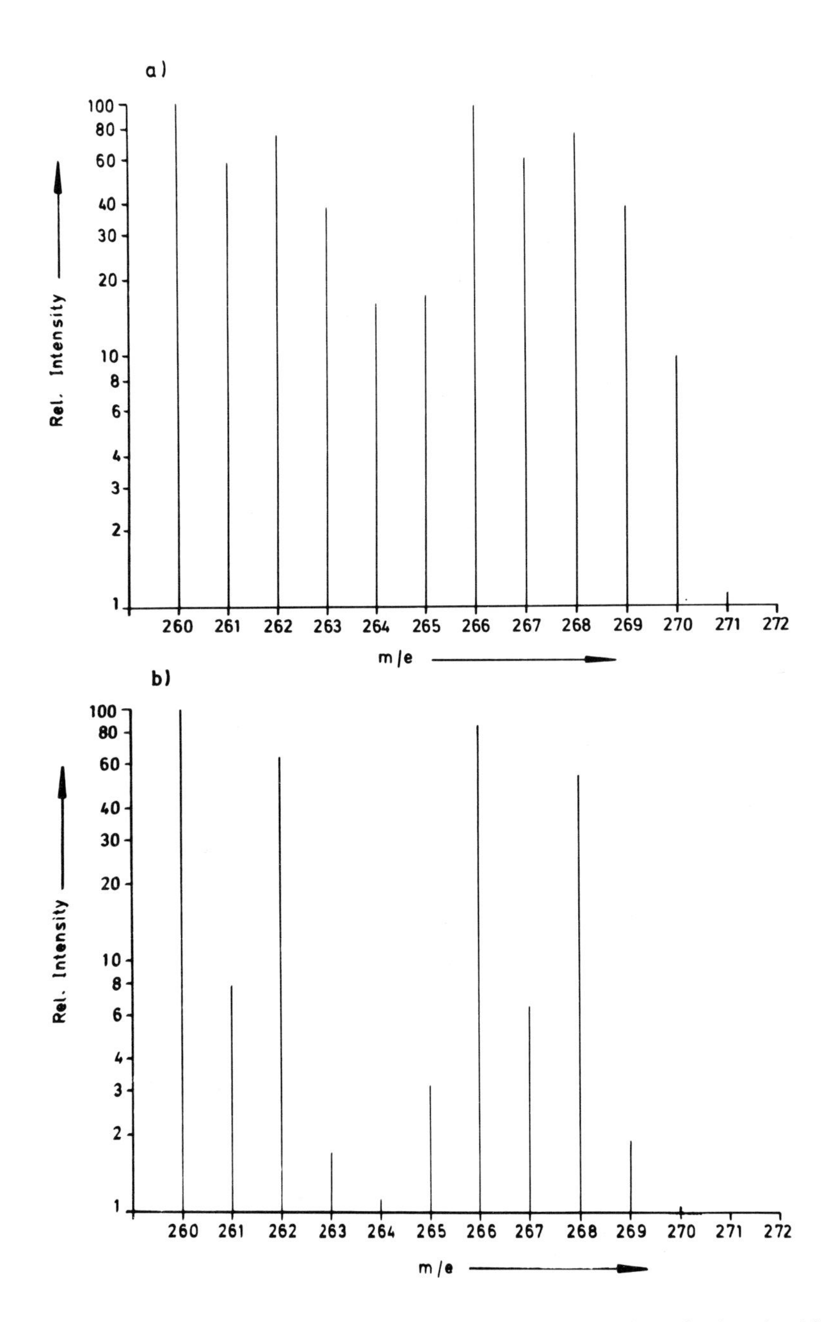

Fig. 2. Field desorption mass spectra of a mixture of cyclophosphamide and cyclophosphamide-d_6 (theoretical molar ratio 1:0.85). (For synthesis of the d_6-analogue see ref. 7.)
(a) Sample amount 1 µg each; (b) sample amount 0.1 µg each. Photographic detection on vacuum-evaporated AgBr photoplates[4].

Table II

Quantitation of mixtures of cyclophosphamide-d_0/-d_6 and -d_0/-d_{10} with *electric detection* by FD MS

The values were obtained by repetitive scan over the molecular ion group and averaging over 30 measurements each. The isotopic distribution of the standard cyclophosphamide-d_{10} was determined by EI MS to be $d_8:d_9:d_{10}$ = 2:18:100. The d_{10} analogue was, like the d_6-analogue, fully C-deuterated in the oxoazaphosphorine ring and was additionally labelled in the alkylating moiety (-N($CD_2CH_2Cl)_2$). The d_{10} analogue was similarly prepared[7] from N,N-bis(2-chloro-1,1-dideuteroethyl)aminophosphorodichloridate[8].

	$\frac{I(m/e\ 260)}{I(m/e\ 266)}$ (theor.)	$\frac{I(m/e\ 260)}{I(m/e\ 266)}$ (exp.)	Deviation from true value (%)
Mixture d_0/d_6	1:0.88	1:0.90 ± 0.04	2.2
	$\frac{I(m/e\ 260)}{I(m/e\ 270)}$ (theor.)	$\frac{I(m/e\ 260)}{I(m/e\ 270)}$ (exp.)	Deviation from true value (%)
Mixture d_0/d_{10}	1:0.77	1:0.81 ± 0.04	5.2

In Fig. 2 is shown the molecular ion group of a mixture of CP-d_0 and -d_6. With the sample amount used for the spectrum in Fig. 2a the resulting densitogram was over exposed with relation to the more abundant isotopic peaks, and therefore quantitative data could not be obtained from these ions. The results with 100 ng of each component shown in Fig. 2b enabled the estimation of the ratio *m/e* 260 : *m/e* 266 to be 1:0.86. To explore the applicability of the stable isotope dilution technique we ran the FD spectra of CP-d_0/-d_6 mixtures of different compositions. The results of this study are shown in Table I and reveal within a range of two orders of magnitude in respect to the relative composition an error of 10% from the true values on average. At high resolution (15,000) 0.1 μg therefore represents a near-optimal value for quantitative determination of CP, but the sensitivity should be enhanced by the use of lower resolutions or different detection systems.

Fig. 3 gives an impression of the good reproducibility of the FD spectrum of CP showing the molecular ion group of a mixture CP-d_0/-d_6 (theoretical molar ratio 1:0.94) of three repetitive scans with electric detection and paper chart recording.

A more detailed study investigating the instrumental precision of the low-resolution instrument used gave the results displayed in Fig. 4. The analysis of these data revealed a coefficient of variance for the mean of 3.7%. Taking into account a confidence range of 3σ for the calculated mean we obtained 0.82 ± 0.09 indicating a confidence of 99.7%. When a larger number of repetitive scans was taken from the molecular ion group of mixtures of CP-d_0/-d_6 and CP-d_0/-d_{10} the accuracy of the FD estimation was between about 2 and 5%.

CONCLUSION

The results described above show that the FD technique yields mass spectra of good reproducibility and allows quantitation in microgram and submicrogram levels. The quantitation of CP was achieved with an error from the true value in the order of 5%. The sensitivity and accuracy of FD MS demonstrate the usefulness of the technique for quantitative biochemical and clinical analysis.

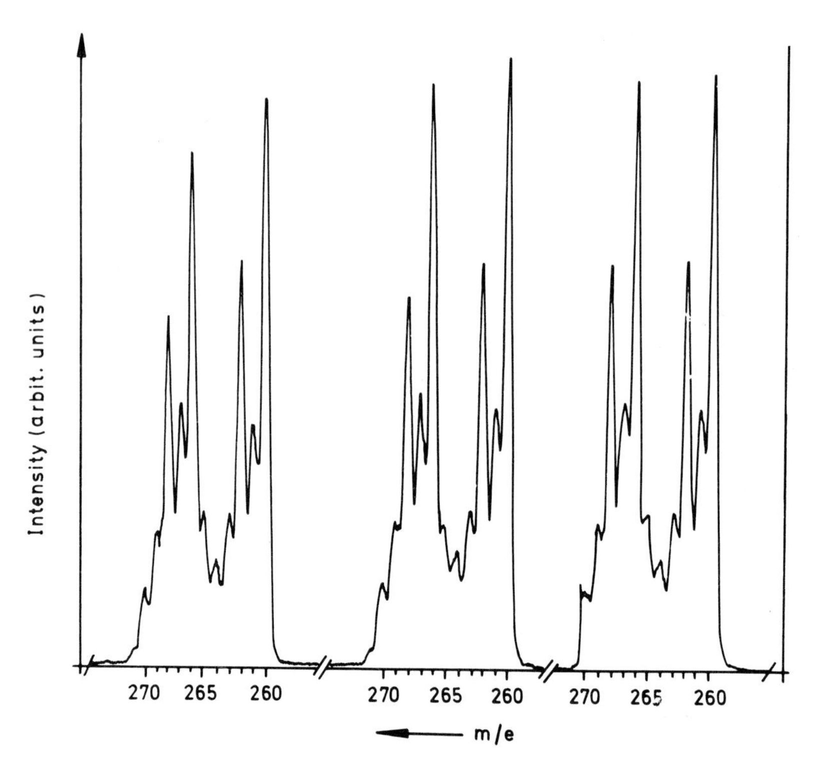

Fig. 3. Section of a set of repetitive scans of the molecular ion group of DP-d_0/-d_6 (low resolution, electric detection).

ACKNOWLEDGEMENTS

This work was supported by grants from the Deutsche Forschungsgemeinschaft, Landesamt für Forschung des Landes Nordrhein-Westfalen and Fonds der Deutschen Chemischen Industrie.

REFERENCES

1 H.D. Beckey and H.-R. Schulten, *Angew. Chem. Int. Ed. Engl.*, 14 (1975) 403.
2 H.D. Beckey and H.-R. Schulten, in T. Pugh and C. Merrit (Editors), *Practical Spectroscopy Series*, Marcel Dekker, New York, in press.
3 H.-R. Schulten, *Methods Biochem. Anal.*, 24 (1977).
4 H.-R. Schulten, *Cancer Treatment Rep.*, 60 (1976) 501.
5 H.D. Beckey, A. Heindrichs and H.U. Winkler, *Int. J. Mass Spectrom. Ion Phys.*, 3 (1970) Appl 9-11.
6 H.-R. Schulten, *Biomed. Mass Spectrom.*, 1 (1974) 223.
7 P.J. Cox, P.B. Farmer, A.B. Foster, E.D. Gilby and M. Jarman, *Cancer Treatment Rep.*, 60 (1976) 483.
8 L.G. Griggs and M. Jarman, *J. Med. Chem.*, 18 (1975) 1102.

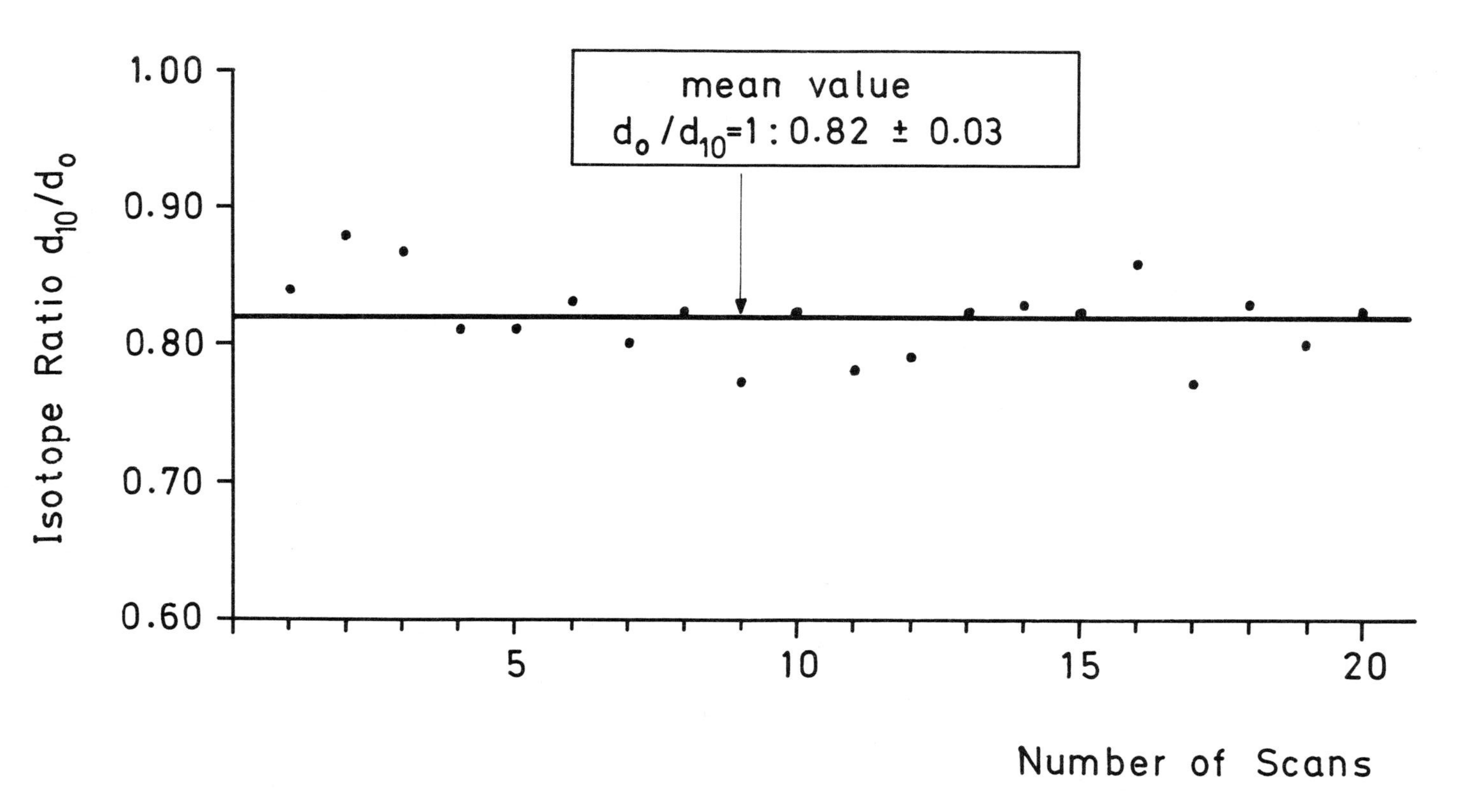

Fig. 4. Instrumental precision for the quantitation of cyclophosphamide by FD MS of a d_0/d_{10} mixture. Low resolution, electrical detection, 0 mA emitter heating current.

THE DESIGN, OPERATION AND APPLICATION OF A LOW-COST ELECTRONIC DEVICE FOR THE DETERMINATION OF ION-INTENSITY RATIOS

A.M. LAWSON, R.J. BULMER, A.E. LOWE and J.F. PICKUP

Divisions of Clinical Chemistry and Bioengineering, Clinical Research Centre, Harrow, Middlesex (Great Britain)

SUMMARY

A low-cost electronic device to monitor two ions and provide the ratio of their intensities is described. The device operates in two modes, repetitive and accumulative. In the repetitive mode consecutive channels are integrated and their ratio displayed and printed, whereas in the accumulative mode, integrals of ion intensities are summed for a period before ratios are given. The unit has been designed principally for application in quantitative experiments using stable-isotope dilution with mass spectrometry. The precision of the ratios generated are demonstrated using hexachlorobutadiene mass ions and a calibration series of mixtures of phosphate and ^{18}O-labelled phosphate as internal standard.

INTRODUCTION

Selected ion monitoring (SIM) mass spectrometry is now a widely used technique for the quantification of compounds. A number of instrumental methods are available for recording individual ion signals and for the collection and assessment of data, *e.g.* for review see ref.1. The data processing aspects are readily handled by computers which can also be used to control the mass spectrometer for selecting and focusing the ions of interest (*e.g.* refs.2 and 3). In general, these systems improve the precision of quantitative measurement, simplify operation and permit the processing of large numbers of samples.

However, many workers make only limited use of selected ion monitoring and many have no dedicated computing facility. In such cases, ion signals are normally recorded by pen or oscillographic recorder and subsequent calculations made by hand. This approach suffers from several disadvantages including the necessary pre-selection of output voltage ranges for each channel and the possibility of making errors of measurement and transcription.

Electronic devices for signal ratioing have been in use in dual collector isotope ratio mass spectrometers for many years, and more recently described[4] in relation to selected ion monitoring and gas chromatography-mass spectrometry. Unlike the steady-state conditions of sample flow in standard isotope ratio instruments, the rapidly changing ion source concentrations of sample compounds, when introduced via a gas chromatograph, impose limitations on the method of signal ratioing and the precision of measurement. The present unit was constructed to monitor two ions and display or print the ratio of their intensities. Important considerations in its design were the requirement of good precision and the capability of measuring ion ratios from both static and dynamic sample concentrations for application in stable-isotope dilution experiments.

EXPERIMENTAL

Instrumentation

The mass spectrometer used for this work was a Varian MAT-731 operated at 1000 RP, 8 KV and 70 eV, with an emission current of 800 µA and source temperature of 250°. Samples can be introduced from a variety of inlets including a small metal reference inlet (220°) with variable leak and a Varian 2700 series gas chromatograph. Mass ion switching was carried out with a standard peak-matching unit (maximum switching, 10% of lowest mass) using accelerating voltage alternation. Peaks are displayed on an oscilloscope screen by the application of a saw-tooth voltage, in phase with accelerating voltage switching, to beam deflection plates placed immediately before the collector and to the horizontal deflection voltage of the display (Fig. 1). This saw-tooth voltage can be varied to display 10, 5, 2 or 1% of the mass scale or a single mass peak on the screen.

Operation and construction of the isotope ratio unit

Signals from ion peaks to be integrated are taken from the secondary electron multiplier (SEM) main amplifier and represent a sweep across either the intact mass peak (Fig. 2a) or the peak maximum (Fig. 2b). The unit can be operated in two modes: a repetitive mode in which the ratio

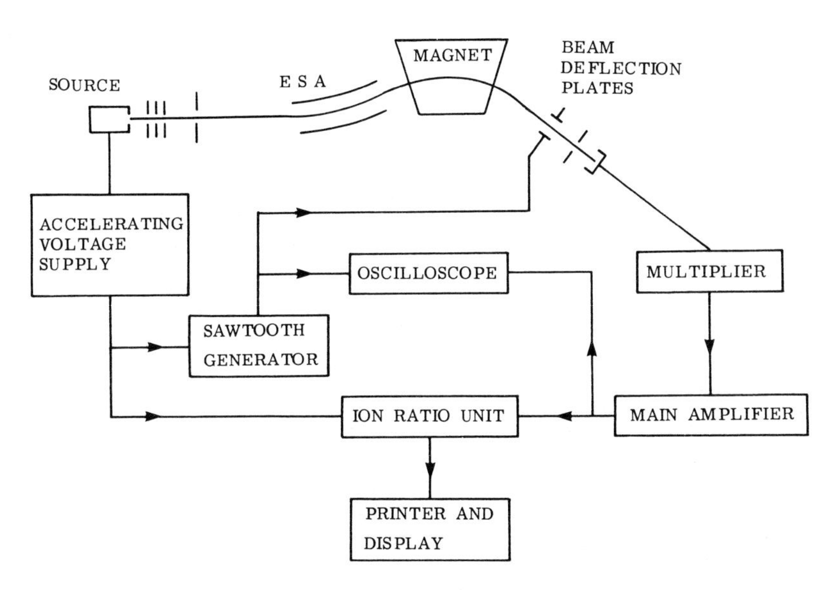

Fig. 1. Schematic diagram of the ion ratio unit in relation to the mass spectrometer.

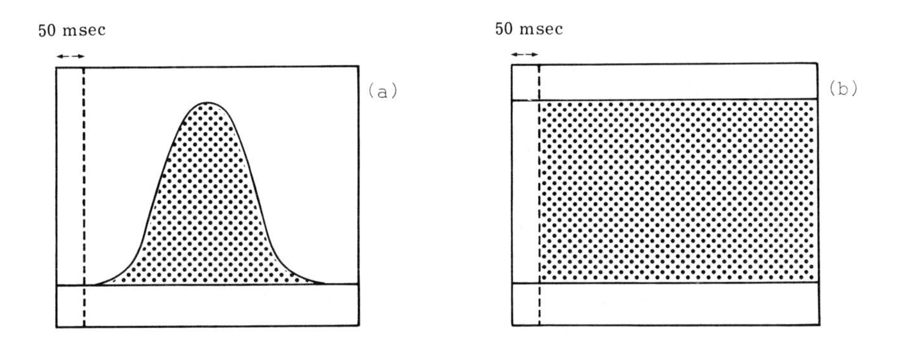

Fig. 2. The signals integrated (dotted area) by (a) scanning across the peak, and (b) by focusing on the peak maximum.

of the intensities is calculated and printed after each pair of sweeps, and an accumulative mode when integration is continued over a number of sweeps until the operator signifies the end of a chromatographic peak of interest.

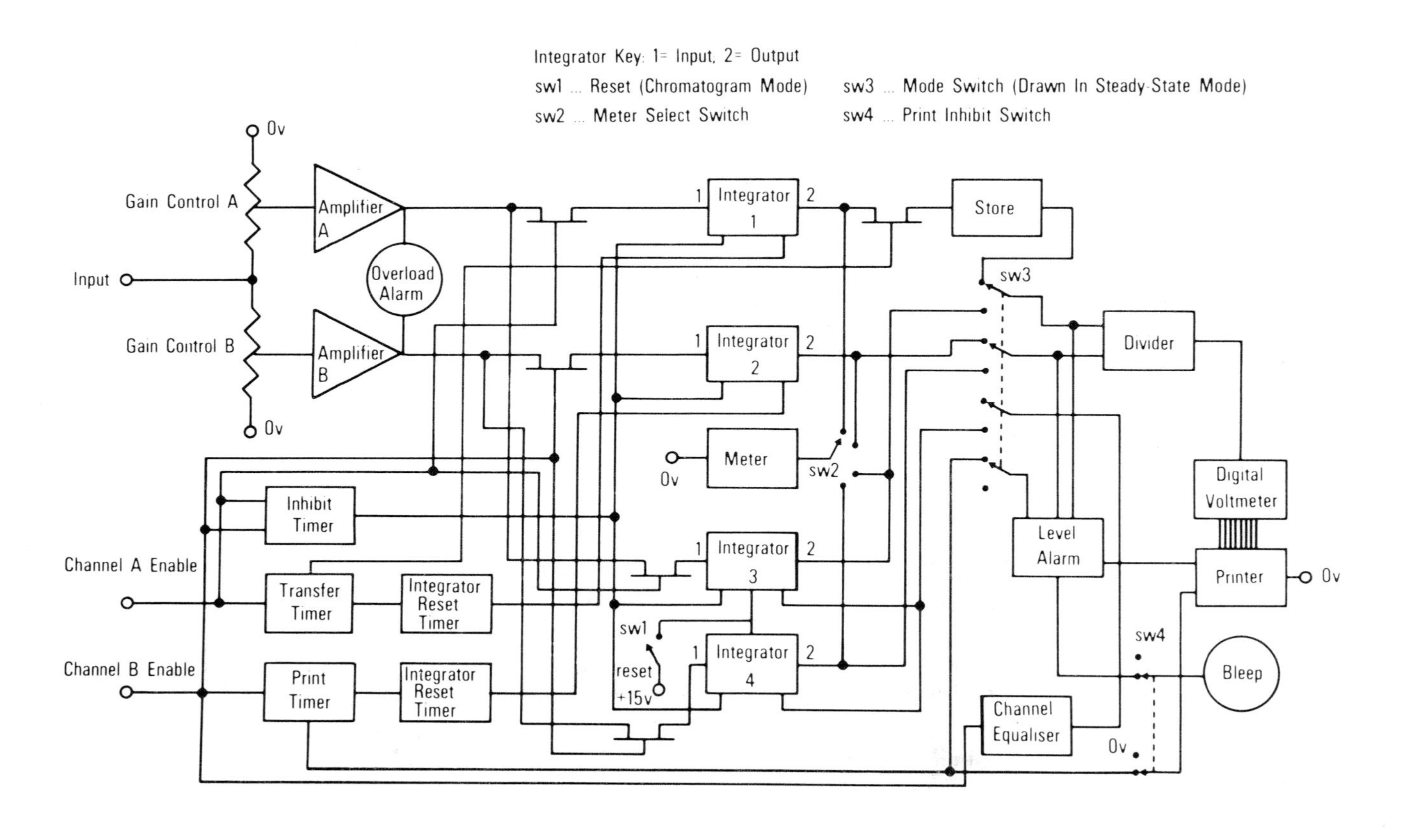

Fig. 3. Schematic diagram of the ion-ratio electronics.

Repetitive mode. A dual output buffer stage has been incorporated into the existing peak-matching unit which extracts the $\bar{Q}$ and $\bar{q}$ signals for external use from the multi-vibrator which drives the relay that switches the accelerating voltage. The 0-24 V switching levels from the dual buffer are in antiphase and designated channel A and channel B in Fig. 3. The ion signal passes to the isotope unit on a single input line and is routed into two independent amplifiers each with variable and accurately known gain. The signals from the two amplifiers are then routed into two separate, accurately matched, integrators, via f.e.t. switches. These switches are commanded by the channel A and B switching signals and are open only when the appropriate channel is at 24 V. The onset of integration is postponed by 50 msec by the inhibit timer, so as to prevent integration of switching transients generated by the changeover between channels.

The operating sequence is as follows. Channel A is energized, opening the path into integrator 1 which accumulates signal. When channel A switches off, the transfer timer gates the output of integrator 1 into a store; shortly after, a reset timer clears integrator 1. Coincident with channel A's switching signal falling to 0 V, channel B goes to 24 V permitting integrator 2 to acquire signal. When channel B switches off the print timer causes the digital voltmeter to sample the output from the divider (which continuously compares the store with integrator 2), and the result is displayed on the DVM and printed. Shortly after this a reset timer resets integrator 2. Coincident with channel B switching off, the new signal passes to integrator 1.

The level alarm monitors the input to the divider continually, and if a print command occurs with either input too high or too low for accurate division, an audible alarm is sounded and a special character is printed to indicate the invalid status of the corresponding printed result.

Accumulative mode. In this mode the input amplifiers are connected to integrators 3 and 4, which have integrating constants approximately 5 times less than 1 and 2, and the ratio is displayed continuously on the DVM. A channel-equalizing circuit ensures that the final result is due to an equal number of readings on each channel. The switch SW1 has three functions: to start integration of signals by integrators 3 and 4, to stop integration and display the final ratio, and to reset the integrators to zero. An additional alarm, in the form of an LED indicator gated from the two outputs of the input amplifiers, lights when either output exceeds 10.5 V. This prevents erroneous results being taken when saturation of the input amplifiers has occurred.

Any standard 3- or 4-digit digital voltmeter can serve to display the ratios. A Weir DVM printer (Model 530) was used to print results in the repetitive mode.

MATERIALS AND METHODS

Hexachlorobuta-1,3-diene (HCBD), spectroscopy grade (BDH, Poole. Great Britain, Lot No. 1418140).

Phosphate standard. Potassium dihydrogen orthophosphate "AnalaR" (BDH, Lot No. 2041610), dried at 150° for 24 h, cooled and stored in a desiccator over silica gel. A standard stock solution (10 mmole/l) was prepared by dissolving 1.3500 g in deionized water and making up to 1 litre. From this solution 5, 4, 3, 2 and 1 mmole/l solutions were prepared.

Labelled phosphate. Disodium hydrogen orthophosphate enriched in ^{18}O (Miles-Yeda, Rehovoth, Israel, Product No. 52-823) certified to contain between 87 and 94% ^{18}O. A stock solution was prepared by dissolving 0.0375 g in 50 ml deionized water.

N-Methyl-N-trimethylsilyl-trifluoroacetamide (MSTFA) (Pierce and Warriner (U.K.), Stockport, Great Britain).

Trimethylchlorosilane (TMCS) (Applied Science Labs., State College, Pa., U.S.A.).

Preparation of phosphate calibration solutions[5]

To 1-ml aliquots of each of the 5, 4, 3, 2 and 1 mmole/l solutions, 1 ml of stock labelled phosphate was added. Each was then passed through an individual 10 cm x 1 cm column of Zeo-Karb 225 cation-exchange resin, 52-100 mesh, ammonium form. Each was eluted with 5 x 1 ml deionized water and the eluates were taken to dryness under nitrogen before derivatization of individual residues with 400 µl MSTFA and 20 µl TMCS.

Ion ratio measurement

HCBD. A sample was introduced from the reference inlet and the ratio of *m/e* 258/260 recorded by the isotope ratio unit in the repetitive mode. This ratio was also measured in the accumulative mode from HCBD introduced from the gas chromatographic (GC)inlet (6 ft. 10% OV-101 glass packed column, 150° isothermal, flow-rate 40 ml/min, retention time 3.5 min). Similarly the ratios *m/e* 261/260 and *m/e* 267/260 were recorded. 1 µg HCBD in ethyl acetate was normally used for GC sampling although smaller mounts (20 ng) were also tested.

Phosphate. The ratio unit, in the accumulative mode, was used to measure the ratios of ions 299 and 307 from accurately prepared mixtures of tris-trimethylsilyl phosphate and a fixed amount of its $^{18}O_4$-labelled internal standard passed through the GC inlet (6 ft. 10% OV-101 glass packed column, 150° isothermal, flow-rate 40 ml/min, retention time 2.3 min). These ions are the respective (M-15) fragments.

RESULTS

HCBD

HCBD was selected as a convenient compound with which to investigate the operation of the isotope ratio unit. A sample was readily volatilized and gave a series of molecular ion peaks with intensities that provided a range of fixed ratios. The theoretical peak heights of HCBD are listed in Table I.

Table I

Theoretical peak heights of HCBD

Ion mass *m/e*	Theoretical peak height[6] (Rel. to *m/e* 260)
258	52.1
259	2.33
260	100
261	4.48
262	79.98
263	3.58
264	34.1
265	1.53
266	8.20
267	0.37
268	1.05
269	0.047
270	0.056

A series of initial tests was carried out to determine the optimal conditions and parameters of both the mass spectrometer and the ratio unit under which ratios could be obtained reliably. Such parameters included the scan time for each channel, amplifier filter values, SEM voltage, level of signal in the integrator before ratioing, signal attenuation, etc. As many factors as possible were held constant throughout the ensuing measurements to test the precision of the unit.

Repetitive mode. Essentially this mode was designed for measuring ion peaks from constant rates of sample flow. The time necessary for the printer to print each result after every two scan restricts the scan time for each channel to 1 sec. Table II lists the values generated for the ratio 258/260 from HCBD introduced from the heated reference inlet.

Table II

Values generated for the ratio 258/260 from HCBD

Mean (n = 33)	0.526
Variance	7.803 x 10^{-7}
Standard deviation	0.000883
Coefficient of variation	0.168%
Theoretical ratio	0.521

Accumulative mode. Channel switching rates of 6 times a second can be used in this mode although sensitivity is reduced owing to the 50 msec inbuilt delay. A time of 0.5 sec for each channel was used for all GC samples.

Table III summarizes the data for the ion ratios 258/260 from HCBD introduced from the gas chromatograph.

Table III

Data for the ion ratios used from HCBD

Ion ratios	258/260		261/260	267/260
Sample size	1 µg	20 ng	1 µg	1 µg
Mean ratio	0.526 (n = 4)	0.525 (n = 17)	0.045 (n = 12)	0.0034 (n = 8)
Standard deviation	0.00053	0.0075	0.002	0.000033
Coefficient of variation (%)	0.1	1.43	0.5	0.97
Theoretical ratio	0.521	0.521	0.043	0.0034

Phosphate

Fig. 4 shows the calibration curve obtained for the trimethylsilyl phosphate mixtures. Each point on the graph represents the mean of four measurements with the exception of the 12 nmole phosphate concentration where eight values were taken, giving a coefficient of variation of 0.25%

for that ratio.

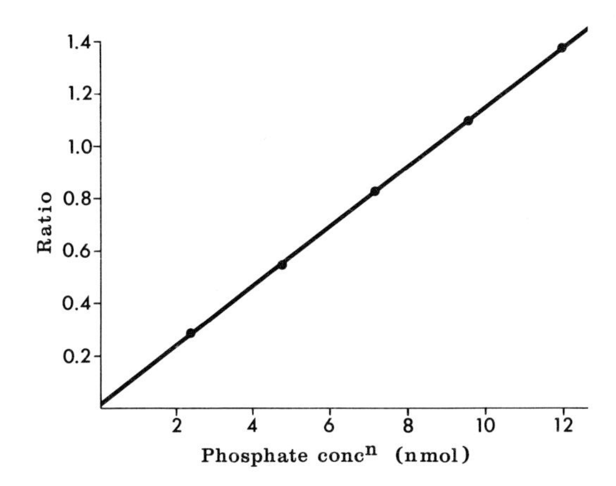

Fig. 4. Calibration curve obtained from measuring the ion ratio *m/e* 299/307 in mixtures of various amounts of phosphate-TMS and a fixed amount (12.5 nmole) of $^{18}O_4$ phosphate-TMS.

DISCUSSION

In quantitative selected ion monitoring mass spectrometry a large number of factors influences the precision with which measurements of ion abundances can be made and these have been discussed elsewhere (refs.1,3, references therein). In the design of the present isotope ratio unit it was decided to concentrate on an approach to obtaining good reproducibility in determining ion intensity ratios and to forego high sensitivity (*i.e.* subnanogram) and a degree of flexibility in the range of assay methods that could be used.

It was projected that the main area of application would be in stable-isotope dilution assays where the stable-isotope abundance was of the same order as the natural material. Low isotopic abundance coupled with very small sample sizes markedly reduces the precision and accuracy achievable in ion intensity measurements[3].

The unit was constructed to operate in conjunction with a peak--matching unit. This proved a convenient means of switching between ions to give a signal that could be integrated either as a complete peak, or as a signal equivalent to the peak maximum for the time period of the scan (Fig. 2). The former option was included to allow for possible drift in the ion focus[4] but proved unnecessary owing to the high stability of the Varian MAT-731 instrument and the use of reduced resolution. For this and sensitivity reasons the peak maximum method was used throughout. An obvious difficulty in determining the accuracy of an isotope ratio measurement is knowing the true natural isotope abundance of the constituent atoms in the ions being measured. Although the mass ions of HCBD should have no contribution from isotopes other than of carbon and chlorine, fractionation of isotopic forms during commercial preparation, or isotope effects in the mass spectrometer giving ion abundances differing from the theoretical values, could not be ruled out. When the abundance values calculated by a probability method [6] were used, the accuracy of measuring the ratio of ions 258/260 from a reference inlet sample and in the repetitive mode was within 1% with a coefficient of variation of about 0.2% (see Table II).

Most biological compounds requiring analysis are present in mixtures. This necessitates the use of GC separation and hence the isotope ratio unit had to be able to cope with GC peaks. The repetitive mode was unsatisfactory in this instance as consecutive channels would be integrated at different sample concentrations. The accumulative mode was designed to overcome this problem by accumulating the integrals of each channel over the entire GC peak. In a peak of 30 sec (base width) each channel is integrated 30 times at a switching rate of 0.5 sec/channel. Table III indicates the high precision obtained for the 258/260 ratio at the level of 1 μg HCBD injected on the gas chromatograph. As expected, the coefficient of variation increased by an order of magnitude when HCBD was injected at the 20-ng level. In a similar way the precision deteriorated for the ratios of ions 261/260 and 267/260 where peaks differing in intensity by approximately 22 and 270 times, respectively, were being ratioed.

To check the application of the unit to the generation of a calibration curve, mixtures of ^{18}O-labelled and natural phosphate were prepared with considerable care and attention to quantitative technique to minimize error. The curve represents a limited concentration range. However, to obtain good precision in a stable-isotope dilution assay it is necessary to use calibration curves that closely encompass the analyte concentration. The phosphate calibration curve was excellent: it is shown in Fig. 4 as a straight line passing near the origin.

The ion ratioing unit described functioned close to its design requirements. It was inexpensive to construct and the precision of ratio measurement was acceptable. The unit should be adequate for most stable-isotope dilution assays of biological compounds where the isotope abundances and the sample concentrations are within the limits discussed.

ACKNOWLEDGEMENTS

We are indebted to Mr. J. Baker for helpful advice and discussion during the work and to the skilled technical assistance of Mr. M.D. Chu and Mr. M.J. Madigan.

REFERENCES

1 F.C. Falkner, B.J. Sweetman and J.T. Watson, *Appl. Spectrosc. Rev.*, 10 (1975) 51.
2 J.F. Holland, C.C. Sweeley, R.E. Thrush, R.E. Teets and M.A. Bieber, *Anal. Chem.*, 45 (1973) 308.
3 R.M. Caproli, W.F. Fies and M.S. Story, *Anal. Chem.*, 46 (1974) 453.
4 P.D. Klein, J.R. Haumann and W.J. Eisler, *Anal. Chem.*, 44 (1972) 490.
5 J.F. Pickup, *Ann. Clin. Biochem.*, 13 (1976) 306.
6 J.F. Pickup and K. McPherson, *Anal. Chem.*, (1976) in press.

A THEORY OF STABLE-ISOTOPE DILUTION MASS SPECTROMETRY

J.F. PICKUP[+]

Division of Clinical Chemistry, Clinical Research Centre, Watford Road, Harrow HA1 3UJ (Great Britain)

and

C.K. McPHERSON[++]

Division of Computing and Statistics, Clinical Research Centre, Watford Road, Harrow HA1 3UJ (Great Britain)

SUMMARY

Among the many published methods of analysis using stable isotope dilution with mass spectrometry there is no agreement as to the relationship between the relative proportions of natural and labelled material and measured isotope ratio. This relationship is often shown as a calibration graph and is described variously by the relationships $y = x$, $y = a + x$, $y = a + bx$, and $y = a + f(x)$. It is argued that satisfactory results are more likely to be obtained from isotope-dilution procedures when results are fitted to the correct equation rather than simply to the equation for a straight line.

Accordingly, this paper sets out the derivation of an equation relating measured isotope ratio to the proportions of labelled and unlabelled material:

[+] Present address: Department of Pharmacy, Royal Cornwall Hospital (Treliske), Truro, Cornwall, Great Britain.

[++] Present address: Department of Social and Community Medicine, 8 Keeble Road, Oxford OX1 3QN, Great Britain.

$$R = \frac{(x/y)(p_1/M) + (q_1/M')}{(x/y)(p_2/M) + (q_2/M')}$$

where R is the ratio of the abundance of the isotopic form representing natural material to the abundance of the isotopic form representing the internal standard (labelled) material; x is the mass of natural material and y is the mass of labelled material in the sample; M is the molecular weight of natural material, and M' is the molecular weight of labelled material; p_1 and q_1 are the probabilities of occurrence of the isotopic form representing natural material in natural and labelled material, respectively; and p_2 and q_2 are the probabilities of occurrence of the isotopic form representing labelled material in natural and labelled material, respectively. R is clearly *not* linearly related to x/y although two special cases are found where the relationship is linear and one of these may be of practical use. Data supporting the validity of the theory are presented.

INTRODUCTION

The recent development of stable isotope dilution as a quantitative technique in biochemistry has been well documented[1]. Presumably the technique is based on extensive earlier use of stable isotopes as internal standards in geochemistry and related fields. A thorough understanding of the theory of stable isotope dilution in the determination of elements exists and may be seen from discussions such as that given by Hintenberger[2]. To what extent this theory is applicable to the determination of molecules has not been demonstrated, and indeed the way in which the analytical response (a measured isotope ratio) is supposed to be related to the relative proportion of labelled and natural material varies from one publication to another.

Thus a calibration graph of measured isotope ratio against relative proportions of natural and labelled material is variously described as a straight line with slope equal to one and intercept zero (*e.g.*, ref. 3), as a straight line with slope not equal to one and a non-zero intercept (*e.g.*, ref. 4), or as a curve of undefined shape (*e.g.*, ref. 5).

Clearly, accurate results are more likely to be obtained if the theoretical equation for the calibration graph is known and is used for standardization. No general theory has been published which relates the measured ratio to the proportions of natural and labelled material;

accordingly, this paper sets out in detail a theoretical basis for stable--isotope dilution assays.

DEVELOPMENT OF THE THEORY

Suppose there are n atoms (or molecules) of natural material existing in i isotopic forms with minimal atomic (or molecular) mass W. (In the case of molecules, isotopically distinct forms having the same nominal mass are considered to be identical - hence the masses form an integral series. Extension of the theory to the high resolution case should not be difficult.)

Similarly let there be m atoms or molecules of labelled (spike) material also existing in the same i isotopic forms. Further, let each isotopic form in each material have an associated probability of occurrence, for the natural material p_j, $\sum_{j=1}^{i}$, and for the labelled material q_j, $\sum_{j=1}^{i} q_j=1$. If the two materials are then mixed, the ratio of any two isotopic forms, a and b, in the mixture will be given by

$$R_{ab} = \frac{np_a + mq_a}{np_b + mq_b} \tag{1}$$

Note that this equation relates the measured isotope ratio R_{ab} to the *numbers* of atoms or molecules. Suppose instead that x grams of natural material and y grams of labelled material are taken. Eqn. 1 can then be rewritten:

$$R_{ab} = \frac{(Ax/M)p_a + (Ay/N)q_a}{(Ax/M)p_b + (Ay/N)q_b} \tag{2}$$

where A is Avogadro's number. M and N are the relative molecular masses of the natural and labelled material, respectively, and are given by:

$$M = \sum_{j=1}^{i}(W + j - 1)p_j \tag{3}$$

and

$$N = \sum_{j=1}^{i}(W + j - 1)q_j \tag{4}$$

where W is the exact relative molecular mass of the isotopic form with minimal mass.

Eqn. 2 may be shortened and rewritten:

$$R_{ab} = \frac{(x/y)\ (p_a/M) + (q_a/N)}{(x/y)\ (p_b/M) + (q_b/N)} \tag{5}$$

Eqn. 5 directly relates the two axes of the usual calibration graph in isotope-dilution studies, R_{ab} and x/y. Clearly, eqn. 5 does not correspond to the general linear relationship $y = a + bx$: the error introduced by assuming that such a relationship exists will depend on the actual values of the parameters in eqn. 5. In particular, it might be that q_a and p_b both tend to zero - there being no interference of natural material at the higher mass or of labelled material at the lower. In this case

$$R_{ab} = \frac{x}{y} \cdot \frac{Np_a}{Mq_b} \tag{6}$$

The relationship is now linear and a calibration graph would be expected to pass through the origin, although the slope of the graph would not necessarily be unity. More commonly, p_b only might be negligible (there being no interference of natural material at higher mass), in which case

$$R_{ab} = \frac{x}{y} \cdot \frac{Np_a}{Mq_b} + \frac{q_a}{q_b} \tag{7}$$

A similar straight line is produced, but with an intercept equal to q_a/q_b.

A SIMPLE TEST OF THE THEORY

A simple test of eqn. 5 was carried out using data given by Bertilsson *et al.*[6]. From the mass spectra of the di-heptafluorobutyl derivative of 5-hydroxyindoleacetic acid given by them, values of p_j and q_j were calculated as follows. If the height of each peak in the published spectrum of natural material is h_j then p_j is found by

$$p_j = \frac{h_j}{\sum_{k=1}^{i} h_k}$$

q_j is found similarly. Note that some error inevitably accrues from this procedure since it is seldom possible to ascribe values to h_k for values of

k above 4 or 5, but while the individual values of h_k and thus p_k are vanishingly small, $\sum_{k=4}^{i} h_k$ may not be insignificant.

Values of M and N were then found by applying eqns. 3 and 4. These values were used to construct a theoretical calibration curve covering the same range of values of x/y as those used used by Bertilsson *et al.* The result is shown in Fig. 1.

From the good agreement between theory and practice it may be concluded that eqn. 5 adequately describes the relationship between the relative proportions of natural and labelled material in a mixture, and the measured isotope ratio.

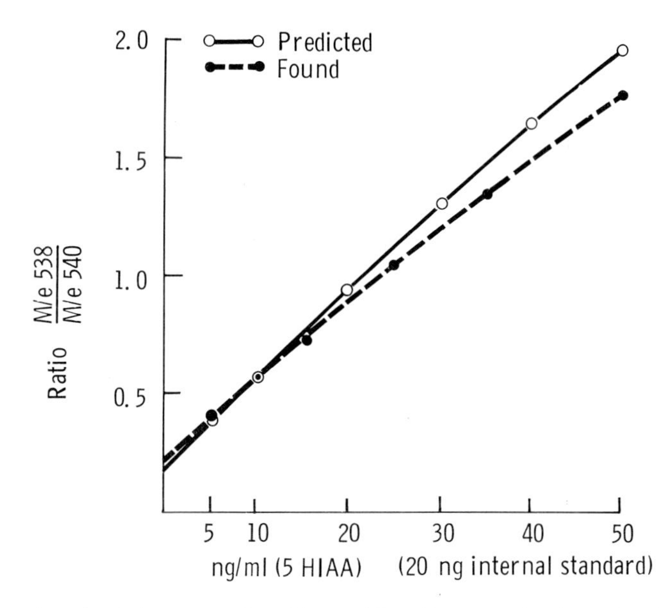

Fig. 1. Data comparing the calibration graph given by Bertilsson *et al.* with the calibration graph calculated from the theory given above.

DISCUSSION

The relationship given by eqn. 5 may be taken to describe the general case of an isotope-dilution experiment. It is apparent that the value of knowing the exact relationship depends on the extent to which an experimentally derived calibration graph departs from linearity. The error that will be introduced by assuming that R_{ab} and x/y are linearly related would not be great in cases such as that shown in Fig. 1, and it may well be that an attempt to fit experimental data to eqn. 5 would not be justified; in fact fitting data such as those given by Bertilsson *et al.* to a quadratic equation ($y = a + bx + cx^2$) could well bring about an improvement in accuracy which could not be bettered by more complex treatment. The real benefit of knowing eqn. 5 is that it indicates the value of attempting to achieve one of the two special cases enumerated in eqns. 6 and 7. Only rarely will it be possible to obtain spike material that contains truly negligible amounts of natural material and thus the case of eqn. 6 is probably of theoretical interest only. On the other hand the case described by eqn. 7 would commonly be realized in practice if the spike material is sufficiently heavily labelled. Typically, an increase in molecular weight of 4 should be sufficient when compounds containing only C, H, O and N are considered.

REFERENCES

1 A.M. Lawson and C.H. Draffan, *Progr. Med. Chem.*, 12 (1975) 1.
2 H. Hintenberger, in M.L. Smith (Editor), *Electromagnetically enriched Isotopes and Mass Spectrometry*, Butterworth, London, 1956, p. 177.
3 M.C. Horning, W.G. Stillwell, J. Nowlin, K. Lertratanangkoon, D. Carrol, I. Dzidic, R.N. Stillwell and E.C. Horning, *J. Chromatogr.*, 91 (1974) 413.
4 W.F. Holmes, W.H. Holland, B.L. Shore, D.M. Bier and W.R. Sherman, *Anal. Chem.*, 45 (1973) 2063.
5 J.E. Holland, C.C. Sweeley, R.E. Thrush, R.E. Teets and M.A. Bieber, *Anal. Chem.*, 45 (1973) 308.
6 L. Bertilsson, A.J. Atkinson, J.R. Althous, A. Harfast, J.E. Lindgren and B. Holmstedt, *Anal. Chem.*, 44 (1972) 143.

COMPARISON BETWEEN THE FLUORIMETRIC AND MASS FRAGMENTOGRAPHIC DETERMINATION OF POLYCYCLIC AROMATIC HYDROCARBONS OF ORGANIC PARTICULATE MATTER

G. BRODDIN and K. VAN CAUWENBERGHE

Chemistry Department, University of Antwerp (U.I.A.), Wilrijk (Belgium)

SUMMARY

A classical procedure for the quantitative determination of polycyclic aromatic hydrocarbons (PAH) uses the fluorimetric technique, which is known to be very sensitive but also to introduce large errors when sensitizing or quenching compounds are present as impurities. Therefore the method requires not only careful isolation of the PAH as a group but also a step that will separate the individual compounds for the fluorimetric determination. These operations are time consuming, and losses will occur during column or thin-layer chromatographic enrichment steps.

Mass fragmentography on the molecular ions of the PAH has been proposed by the authors as an alternative for quantification. Experimentally, highly reproducible results were obtained by direct injection of aerosol extracts without preliminary separations. Interferences by other compounds were not observed. The only limitation was gas-chromatographic resolving power, which was insufficient for the separation of certain isomeric pairs.

From the several separation systems described in the literature for the isolation of PAH, a choice was made. Experiments confirmed that two-dimensional thin-layer chromatographic separation of the neutral organic compounds of particulate matter, as proposed recently by Herce and Katz, was a very elegant method for the isolation step preceding fluorimetry.

In a synthetic mixture containing PAH with four, five and six rings at least six compounds can be easily separated and measured: benzo(*e*)pyrene, benzo(*a*)pyrene, benzo(*k*)fluoranthene, perylene (molecular ion at mass 252), benzo(*ghi*)perylene and *o*-phenylenepyrene (molecular ion at mass 276).

With the exception of the benzopyrene isomers that have to be measured together, all compounds can be determined by mass fragmentography using quaterphenyl as an internal standard (mass 306).

Results are expected to be different for artificial mixtures with known composition and for natural samples of particulate matter where the large number of compounds may create more problems of interference for the fluorimetric method. Also desorption from thin-layer plates tends to lower the results from fluorimetry more than with mass fragmentography.

INTRODUCTION

In the last few years, the quantitative determination of polycyclic aromatic hydrocarbons (short name PAH) in environmental samples has gained new interest because of the health hazards caused by the presence of these more or less carcinogenic compounds on the particulates of the atmosphere[1]. Although the measurement of benzo(*a*)pyrene as the most active carcinogenic aromatic hydrocarbon may be questionable from the point of view of its chemical reactivity, many authors have proposed analytical procedures that focus on the isolation of this compound before determination is attempted. Column, thin-layer, and paper chromatography are the separation methods used, whereas UV spectrophotometry, fluorimetry and gas chromatography (GC) allow measurement of the isolated compound[2-4].

EXPERIMENTAL

In our laboratory, experience with natural samples has shown that the most typical compounds of this class are the five- and six-ring polyaromatics[5,6] shown in Table I.

Among these compounds benzo(*b*)- and benzo(*j*)fluoranthene are not available as pure reference compounds.

Mass fragmentography on the molecular ions of the PAH, using quaterphenyl as an internal standard, has been proposed by the authors as an alternative procedure for quantification. Therefore the mass spectrometer has to be focused on the masses 252 and 276 for the unknowns and on mass 306 for the internal standard. Highly reproducible results (5% relative standard deviation) were obtained by direct injection of aerosol extracts without preliminary separation, corresponding to nanogram quantities on the column. There were no interferences in this method. The original GC resolving power proved to be insufficient for the separation

Table I

Compounds investigated

Compound	Abbreviation	Mol. weight
Benzo(*a*)pyrene	BaP	252
Benzo(*e*)pyrene	BeP	252
Benzo(*k*)fluoranthene	BkF	252
Benzo(*b*)fluoranthene	Bbf	252
Benzo(*j*)fluoranthene	BjF	252
Perylene	Pe	252
o-Phenylenepyrene (indeno(123, *cd*)pyrene)	o-pheP	276
Benzo(*ghi*)perylene	BghiPe	276

of the isomeric pair benzo(*a*)pyrene and benzo(*e*)pyrene. The use of a slightly more polar column (OV-17 instead of OV-1 or Dexsil 300) resulted in a 50% valley separation of both isomers, enabling their measurement as individual compounds. The pattern of the GC separation is shown in Fig. 1.

To evaluate the accuracy of the method, a comparison of results was attempted with those obtained by an independent analytical procedure, the fluorimetric technique. On the occasion of an inter-comparison test on

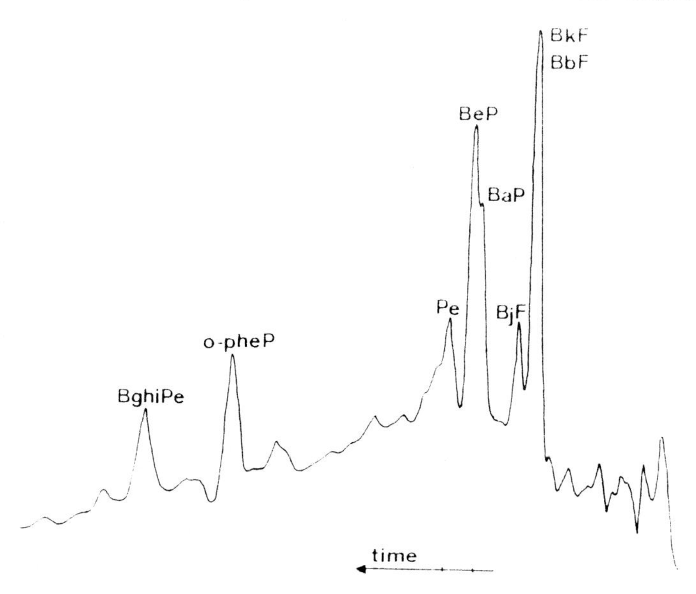

Fig. 1. GC pattern for the 5- and 6-ring polyaromatics.

a sample of Milan dust, the response of different laboratories was very limited. Results involving a GC approach proved to be relatively close but fluorimetric determinations were unreliable (Table II).

These data stimulated us to look for a separation procedure in the literature that would result in the isolation of the above-mentioned individual compounds and to measure them by the fluorimetric method. Such values can be compared with the mass-fragmentographic results after they have been corrected for desorption losses from the thin-layer plate. In the absence of sensitizing or quenching compounds, present as impurities in the adsorbent, the compounds or the solvents used, a reasonable agreement should be obtained between the two methods.

Experiments confirmed the results obtained by Pierce and Katz[7] for the separation of the 5- and 6-ring PAH of major importance. These authors used two consecutive thin-layer chromatographic steps: the first on a aluminium oxide to isolate three isomeric groups of PAH, the second step used 40% acetylated cellulose for the resolution of the individual compounds. Both separations can be favourably combined into a two--dimensional thin-layer chromatographic procedure as shown in Fig. 2.

The R_F values obtained for the individual PAH show good reproducibility (Table III).

Suspended particulate matter was collected by filtration over glass fiber at a flow-rate of 25 m^3/h for periods of 72 h. Quantitative yields of the organic compounds were obtained after 8 h extraction with benzene in a Soxhlet apparatus. The extract was then separated into 3 groups

Table II

Results for Milan dust samples

Compound	Mass fragmentography	Gas chromatography[+]
Benzofluoranthenes	75[++]	116
Benzo(*a*)pyrene	30	29
Benzo(*e*)pyrene	20	15
Perylene	7	15
o-Phenylenepyrene	25	20
Benzo(*ghi*)perylene	24	20

[+] Liberti *et al.*, determination after TLC separation.

[++] Results expressed in ppm weight (microgram per gram dust).

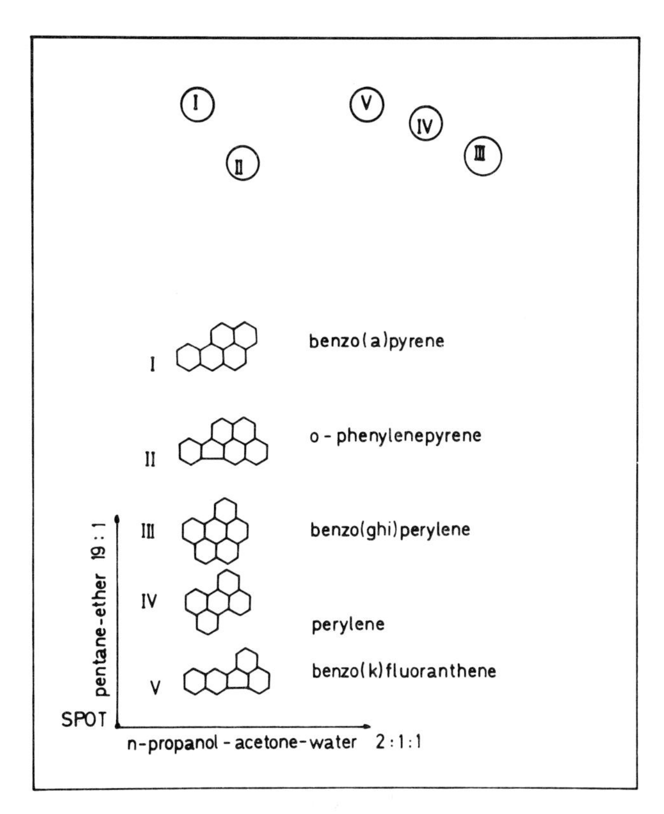

Fig. 2. Two-dimensional thin-layer chromatographic separation of 5 PAH on an alumina-30% acetylated cellulose (1:1) plate.

Table III

R_F values for some polycyclic aromatic hydrocarbons

Compound	Alumina with *n*-pentane-ether (19:1)	30% acetylated cellulose with *n*-propanol-acetone-water (2:1:1)
Benzo(*a*)pyrene	0.85 ± 0.02	0.18 ± 0.02
Benzo(*k*)fluoranthene	0.87 ± 0.02	0.43 ± 0.01
Perylene	0.86 ± 0.02	0.54 ± 0.02
o-Phenylenepyrene	0.80 ± 0.02	0.29 ± 0.03
Benzo(*ghi*)perylene	0.79 ± 0.02	0.67 ± 0.01

according to the scheme, proposed by Hueper *et al.*[8], shown in Fig. 3. After thin-layer resolution of the individual compounds the perimeters of the fluorescent areas for these individual compounds from both the natural samples and the synthetic mixture were scribed, and the adsorbent was removed from the plate. Elution was performed with diethyl ether. After solvent evaporation each residu was taken up into a specific volume of spectrograde cyclohexane for the fluorescence analysis. Spectroscopic data for both analytical procedures are summarized in Table IV. Excitation and emission wavelengths used in this study were checked experimentally on a spectrofluorimeter with double monochromators.

From the fluorescence measurement of individual compounds before and after the thin-layer separation procedure of the synthetic mixture the recovery percentage can be calculated for the PAH available, as is shown in Table V.

In two natural 72 h samples, six polycyclic aromatic compounds were analyzed by the above-described analytical methods. Values obtained by

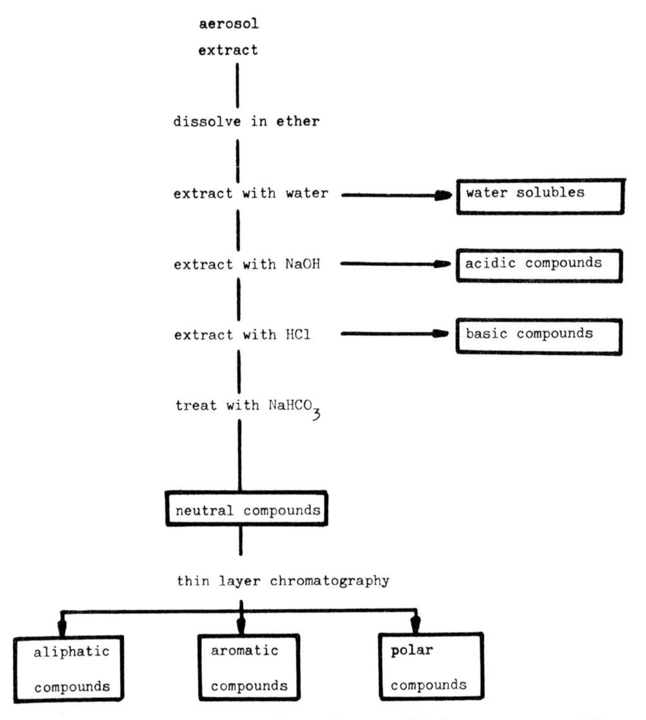

Fig. 3. Huepers scheme for the preliminary separation of the organic fraction of an aerosol extract.

Table IV

Fluorimetric emission and excitation maxima and selected ion masses of some PAH

Compound	Emission maximum (in cyclohexane) (nm)	Excitation maxi- mum) (in cyclo- hexane) (nm)	Selected ion mass
Benzo(*a*)pyrene	380	405	252
Benzo(*k*)fluoranthene	309	402	252
Perylene	409	439	252
o-Phenylenepyrene	360	499	276
Benzo(*ghi*)perylene	383	421	276

Table V

Recoveries from two-dimensional TLC separation

Compound	Run				Average (%)
	1	2	3	4	
Benzo(*a*)pyrene	82	90	84	86	86 ± 4
Benzo(*k*)fluoranthene	86	92	90	88	89 ± 3
Perylene	86	84	90	86	87 ± 3
o-Phenylenepyrene	78	84	87	83	83 ± 5
Benzo(*ghi*)perylene	86	82	84	80	83 ± 3

fluorimetry were corrected for desorption losses in correspondence with Table V. The results are presented in Table VI.

CONCLUSIONS

From these data a number of conclusions can be made.

1. The agreement between the fluorescence and mass fragmentographic results is very reasonable. Relative standard deviations on mass fragmentography values varied between 3 and 8%. With fluorescence measurements the major source of error was the correction for incomplete desorption from the TLC plate. The precision in the fluorimetric determination was normally somewhat better than in GC-MS, ranging from 2 to 6%.

Table VI

Comparison of two natural samples

Compound	Fluorescence	Mass fragmentography
Sample 30/4/76		
Benzo(*a*)pyrene	3.23+	3.39
Benzo(*e*)pyrene	-	2.47....5.86
Benzo(*k*)fluoranthene	2.90	6.56++
Perylene	0.78	0.90
o-Phenylenepyrene	5.85	5.37
Benzo(*ghi*)perylene	3.99	3.72
Sample 7/5/76		
Benzo(*a*)pyrene	3.36+	2.90
Benzo(*e*)pyrene	-	1.82....4.72
Benzo(*k*)fluoranthene	3.38	5.50++
Perylene	0.86	0.80
o-Phenylenepyrene	2.18	1.42
Benzo(*ghi*)perylene	2.38	1.74

\+ Results expressed in micrograms per 1000 m^3 of air.

++ Sum of benzo(*k*)- and benzo(*b*)fluoranthene is measured.

No measurement was performed on benzo(*e*)pyrene by fluorimetry. Mass fragmentographic values for this compound were obtained by a perpendicular drop integration technique.

Two isomers of benzo(*k*)fluoranthene - benzo(*b*)- and benzo(*j*)fluoranthene - were not available as reference compounds. Also, benzo(*b*)fluoranthene has a retention time almost identical with that of benzo(*k*)fluoranthene. Therefore mass fragmentography values represent the sum of *k*, *b* and *j* isomers, whereas fluorimetric values correspond to the *k* isomer only.

2. No evidence could be found for sensitizing or quenching interactions in the fluorescence measurements providing adsorbents and solvents were thoroughly purified before use.

3. Both mass fragmentographic and fluorimetric quantification of PAH are very sensitive. In our experiments, nanogram amounts on column as well as on the TLC plate could easily be measured.

4. Finally, analysis time for both methods can be competitive. Calibration curves in mass fragmentography are more time dependent than in fluorimetry. Therefore, a thin-layer separation can be performed during the time necessary to recalibrate the mass spectrometer for quantitative work.

ACKNOWLEDGEMENT

This investigation was part of the Belgian National Environmental Air Research Program supported by the Belgian Government.

REFERENCES

1 D. Hoffman and E.L. Wynder, in A.C. Stern (editor), *Air Pollution*, Vol. 2, Academic Press, New York, 2nd ed., 1968, Ch. 20, pp. 197-199.
2 E. Sawicki, *Chemist Analyst*, 53 (1964) 24.
3 E. Sawicki, *Chemist Analyst*, 53 (1964) 56.
4 E. Sawicki, *Chemist Analyst*, 53 (1964) 88.
5 W. Cautreels and K. Van Cauwenberghe, *Atmos. Environ.*, 10 (1976) 447.
6 W. Cautreels and K. Van Cauwenberghe, *J. Chromatogr.*, submitted for publication.
7 R.C. Pierce and M. Katz, *Anal. Chem.*, 47 (1975) 1743.
8 W.C. Hueper, P. Kotin, E.C. Tabor, W.W. Payne, H. Talk and E. Sawicki, *Arch. Pathol.*, 74 (1962) 89.

COMPUTERIZED ROUTINE MASS FRAGMENTOGRAPHY OF SOME TRANSMITTER METABOLITES

CLAES-GÖRAN FRI

Application Laboratory, LKB Producter AB, Stockholm-Bromma (Sweden)

A computerized system for routine mass fragmentography or selected--ion monitoring in combined gas chromatography and mass spectrometry using a magnetic sector instrument (LKB 2091) will be described. The system consists of a disk-oriented data system capable of measuring up to 12 selected ions with various times (0.1 sec to 3200 sec) and dwell periods (2 msec to 3.2 sec). Calculation and adjustment of accelerating voltages for each selected ion are done automatically. System drift can be corrected by using one channel as a reference channel to correct all other channel voltage settings. The voltage of each channel can also be modulated so that maximal intensity readings are ensured. The accuracy is 14 bits and the dynamic range is 16 bits. Routine results from this system will be shown for some transmitter metabolites.

DETERMINATION OF THE STRUCTURE OF ORGANIC PEROXIDES BY MASS SPECTROMETRY

T. KASTELIC-SUHADOLC and V. KRAMER

Biochemical Institute of the Medical Faculty, Ljubljana, and Institute Josef Stefan, Ljubljana (Yugoslavia)

Peroxides are known to be present in biochemical materials. Their identification and analysis are difficult because they appear as intermediates in minute quantities only. In spite of their thermal instability, mass spectrometry might be a suitable method for their structure determination and analysis. We tested the method of mass spectrometry with three different types of peroxide: dihydroperoxy-dialkyl ethers, dihydroperoxy-dialkyl peroxides and cyclic peroxides.

The mass spectra of dihydroperoxy-dialkyl peroxides and cyclic peroxides showed molecular ions. The structure of peroxides that do not show molecular ions, such as dihydroperoxy-dialkyl ethers, was determined from the intense fragment ions, which are characteristic for this type of structure. The most stable dihydroperoxy-dialkyl ether, 4,4'-dihydroperoxy-4,4'-diheptyl ether, showed the molecular ion in its mass spectrum.

To obtain usable mass spectra of the above peroxides their properties must be considered.

QUANTITATIVE EVALUATION OF CARBOHYDRATE METABOLISM AT REST AND IN PHYSICAL EXERCISE BY THE USE OF NATURALLY LABELED ^{13}C-GLUCOSE

F. MOSORA[+], P. LEFEBVRE[++], F. PIRNAY[+++], M. LACROIX[+], A. LUYCKX[++] and J. DUCHESNE

[+]*Department of Atomic and Molecular Physics, University of Liège, Sart-Tilman (4000 Belgium)*
[++]*Division of Diabetes, Institute of Medicine, University of Liège (4000 Belgium)*
[+++]*Service of Social Medicine, Institute Malvoz, Liège (4000 Belgium)*

SUMMARY

As demonstrated by Duchesne and co-workers, the $^{13}C/^{12}C$ ratio in the air breathed out by higher animals depends on the isotopic composition of the food.

Because of their particular photosynthetic pathway, some vegetables, *e.g.* maize and sugar cane, are slightly richer in ^{13}C than are most common foodstuffs. This weak but significant enrichment (0.02%) allows the use of maize glucose as a natural and non-radioactive tracer by simply measuring, by double collector mass spectrometry, variations of the $^{13}C/^{12}C$ ratio of the CO_2 exhaled. This is therefore a convenient method for following the complete conversion of an oral glucose load into CO_2 in both animals and man.

Simultaneous measurements of the $^{13}C/^{12}C$ ratio and of the volume of exhaled CO_2 permit us to obtain quantitative evaluation of carbohydrate metabolism for normal subjects at rest or in exercise. We observed that about one-third of a 100-g glucose load is oxidized to CO_2 in 7 h at rest, whereas the whole of the exogenous glucose is metabolized after 4 h of physical activity (half max. O_2 consumption).

Comparisons are made with results from other methods and the advantage of this new and inocuous procedure is emphasized.

INTRODUCTION

Some plants, such as maize and sugar cane, are slightly richer in ^{13}C than are most vegetables used as foodstuffs[1]. Having a dicarboxylic acid pathway preceding the Calvin cycle, the photosynthetic anabolism of these plants incorporates carbon from atmospheric CO_2 with only one ^{12}C enrichment stage in place of the two that exist in the majority of plants. Therefore, the carbon content of these particular plants is less enriched in ^{12}C and all the derivatives appear as "naturally labeled in ^{13}C".

Knowing that the $^{13}C/^{12}C$ ratio in the air breathed out by higher animals depends on the isotopic composition of the food[2,3], we can use the glucose extracted from maize[3-5] as a particularly good tracer to study carbohydrate metabolism.

This report presents quantitative results obtained by simultaneous measurements of the $^{13}C/^{12}C$ ratio and of the volume of exhaled CO_2 during the metabolism of an oral load of "naturally labeled ^{13}C-glucose" in normal subjects at rest or in physical exercise.

A comparison is made with the results obtained from other methods and the originality and usefulness of this new quantitative procedure is outlined.

MATERIALS AND METHODS

Four normal volunteers, aged 21-25 years, were investigated. All were of normal body weight according to standard criteria[4] and none had a family history of diabetes or postprandial glycosuria. In a preliminary test, all were proven to have a normal 100-g glucose tolerance and no glycosuria exceeding 200 mg within 5 h following the glucose load. All the volunteers gave informed consent to this study.

The subjects, fasted overnight, received 100 g of maize glucose (Glucopur, Glucoseries Réunies, Alost, Belgium) in 400 ml of water and were tested during a 7-h period at rest, or during a 4-h period of physical exercise.

The physical exercise consisted of a walk on a treadmill at a speed of about 4.5 km/h with a 10% upward slope and corresponded to a rate of metabolism of about half the maximal individual oxygen consumption.

During both tests, the air breathed was sampled to determine the $^{13}C/^{12}C$ ratio and to measure the amount of exhaled CO_2.

Determination of the $^{13}C/^{12}C$ ratio

Samples of expired air were collected into 1-litre rubber balloons, two samples at time 0 (ingestion of glucose), one every hour during the next 7 h at rest and one every half-hour during the next 4 h of physical exercise. CO_2 and water were immediately separated from the air by trapping in liquid nitrogen and using a vacuum pump. The CO_2 was then evaporated into the mass spectrometer while the water remained trapped in a mixture of methanol and dry ice. The same procedure was used when a sample of the glucose used for the load was converted into CO_2 by burning to determine its ^{13}C content. To estimate the very small differences in the isotopic ratio of the samples (maize has a ^{13}C content of 1.10 atom % while common food has 1.08 atom %) we used the very sensitive method of double collector mass spectrometry which compares the ratio of peak heights of *m/e* 45 and *m/e* 44 of the sample studied with that of a CO_2 standard. The mass spectrometer used was the Varia MAT CH5, and the standard was the graphite sample 21 of the National Bureau of Standards (NBS21). Our results are expressed as $\delta\ ^{13}C$ using Craig's formula:

$$\delta\ ^{13}C = 10^3 \cdot \left[\frac{(^{13}C/^{12}C)_{sample}}{(^{13}C/^{12}C)_{NBS21}} - 1 \right]$$

and they can be related to the PDB scale by subtracting the constant amount 28 (ref. 6). One unit of $\delta\ ^{13}C$ corresponds to a change of about 1.10^{-5} in the ratio $^{13}C/^{12}C$, and the precision of a measurement is 0.2.

Determination of the amount of exhaled CO_2 (V_{CO_2})

Each half-hour, after two minutes of adaptation, the subjects breathed through a double low-resistance valve for a period of 5 min, and the total amount of exhaled air was collected in a Douglas bag and measured by a Tissot gasometer. After water vapor had been trapped, an aliquot of this collected air was analyzed for its CO_2 content in a respiration mass spectrometer Varian MAT M3, calibrated before each determination of sample with two air standards of accurately known composition.

The external glucose consumed during each period (hour or half hour) following the ingestion of the load was expressed by calculating the weight of glucose (1 mole of CO_2 (22.2 l) comes from 1/6 mole of glucose (*i.e.* 30 g) corresponding to the fraction of the expired CO_2 that arises from the oxidation of the ingested glucose. This fraction was calculated by multiplying the total amount (in litres) of CO_2 expired during the period (hour or half hour, see Table IB) by the percentage of the exogenous carbon contribution as determined by the $\delta\,^{13}C$ measurements. If the whole of the exhaled CO_2 came from the metabolism of the maize glucose load, its $\delta\,^{13}C$ would have, neglecting the small isotopic fractionations in the metabolism[3-5], the same value as the substrate, *i.e.* 17.8 ± 0.3 (mean ± S.D., 6 determinations). Therefore, the difference between 17.8 and the $\delta\,^{13}C$ of the CO_2 samples collected at time 0, which is about 3.7 (see Table IA), corresponds to a 100% contribution of the exogenous glucose to CO_2 production. The difference between $\delta\,^{13}C$ at the *n*th hour or *n*th half-hour and the initial $\delta\,^{13}C$ divided by the preceding difference (about 14) gives the percentage of external glucose contributing to the CO_2 at this moment.

RESULTS

Table I summarizes the experimental values of $\delta\,^{13}C$ and V_{CO_2} (mean value for each minute of the period) obtained for each subject at rest as well as in physical exercise.

After the administration of naturally labeled ^{13}C-glucose the $\delta\,^{13}C$ of expired CO_2 rose significantly, with a maximum after 1-1.5 h in the case of physical exercise. At rest this maximum was shifted to between the third and fourth hours. The return to the initial value occurred after 4 h (physical exercise) or 7 h (at rest)[4,10]. The V_{CO_2} slightly increased after glucose ingestion and in both cases began to decline at the second hour.

Table II shows the individual values of exogenous glucose oxidized as calculated from determinations of $\delta\,^{13}C$ and V_{CO_2}.

The total amount of exogenous glucose completely oxidized to CO_2 averaged 29.6 ± 1.0 g (mean ± S.E.M.) in 7 h for the subjects at rest whereas during physical exercise almost the total load given was metabolized after 4 h, *i.e.* 94.8 ± 1.6 g (mean ± S.E.M.). These values represent about 30 and 100%, respectively, of the load given.

The kinetics of the mean amount of glucose oxidized (Fig. 1) followed the $\delta\,^{13}C$ variations. The utilization of exogenous glucose began in physical exercise at the first half-hour and reached a maximal rate of 40 g/h between the first and second hours, whereas at rest the maximal rate was lower (6 g/h) and occurred between the third and fourth hours.

Table I

Experimental data of expired air after oral glucose ingestion

A. $\delta^{13}C$ *determinations*

At rest					In physical exercise				
Time (h)	C	Le	Lo	G	Time (h)	C	Le	Lo	G
0	4.4	3.3	3.2	3.4	0	3.9	3.6	2.9	4.6
1	6.9	5.4	5.2	5.5	0.5	6.5	5.3	5.4	7.1
2	7.4	7.2	6.9	6.9	1	7.2	7.5	7.2	10.0
3	8.7	8.6	7.5	7.8	1.5	7.5	7.4	6.9	8.8
4	9.1	8.0	8.2	7.4	2	6.1	7.6	6.1	6.8
5	7.0	8.1	6.5	6.7	2.5	5.5	8.0	6.8	6.0
6	6.7	7.2	5.2	6.0	3	4.2	6.8	4.6	5.7
7	5.4	6.3	5.1	5.5	3.5	4.4	5.0	4.1	5.3
					4	5.2	5.1	NM[+]	5.0

B. V_{CO_2} *(cm*^{3}*/min)*

At rest					In physical exercise				
Time (h)	C	Le	Lo	G	Time (h)	C	Le	Lo	G
0	283	196	216	224	0	1910	1425	1734	1708
1	252	182	285	199	0.5	2143	1486	1718	1790
2	257	235	256	276	1	2082	1452	1793	1872
3	288	200	226	256	1.5	2080	1441	1755	1833
4	283	195	236	234	2	2105	1430	1746	1803
5	309	185	201	203	2.5	1992	1480	1464	1731
6	283	215	229	217	3	1927	1419	1484	1674
7	239	156	230	243	3.5	1838	1393	1414	1691
					4	1937	1334	NM[+]	1625

[+]NM, not measured.

Table II

Consumption of exogenous glucose

At rest (g/h)					In physical exercise (g/30 min)				
Time	C	Le	Lo	G	Time	C	Le	Lo	G
1	1.9	1.1	1.4	1.3	0.5	7.1	3.5	2.8	4.3
2	4.2	3.3	4.6	3.7	1	19.5	11.9	16.7	23.7
3	5.9	6.2	5.8	6.2	1.5	21.0	15.9	20.2	27.6
4	7.2	5.7	5.9	5.8	2	17.7	16.2	17.3	17.8
5	6.6	4.8	5.1	4.5	2.5	11.2	17.6	15.6	9.8
6	4.3	4.7	3.2	3.6	3	5.1	15.9	11.3	6.9
7	2.5	3.4	2.4	3.1	3.5	4.0	10.6	5.8	4.8
					4	5.0	5.9	NM[+]	2.0
Total	32.6	29.2	28.4	28.2	Total	90.6	97.5	89.7[++]	96.9

[+] NM, not measured.

[++] Total for 3.5 h.

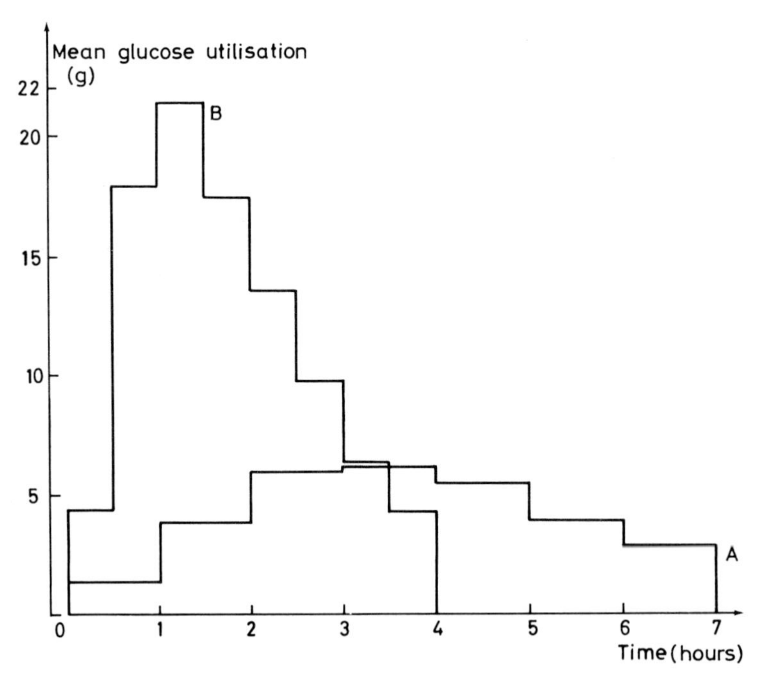

Fig. 1. Consumption of exogenous maize glucose (calculated from $\delta\,^{13}C$ of the expired CO_2 and from V_{CO_2}) for 4 subjects: at rest (A) and in physical exercise (B).

DISCUSSION

Our simultaneous determinations of $\delta\ ^{13}C$ and V_{CO_2} permit calculations, with reasonable precision (better than 10%), of the amount of exogenous glucose oxidized to CO_2 during rest or exercise. At rest, two-thirds of the load was stored and only one-third was consumed during a period of 7 h; during exercise the whole load was oxidized in 4 h.

In both cases, the external load of glucose contributed less than one-third of the total CO_2 exhaled: we observed, in fact, that the maximal value of $\delta\ ^{13}C$ was around 9 and not 17.8 (the $\delta\ ^{13}C$ of the glucose given).

If we compare our results with those recently obtained, with more difficulty, by ^{14}C measurements[7], with substances artificially ^{13}C labeled, or by consideration of the glucose levels in arterial blood flowing out of the splanchnic bed[9], we find a good agreement between these various methods.

On the other hand, classical measurements of the respiratory quotient (RQ) and of blood glucose (venous glycemia) were carried out simultaneously with our $\delta\ ^{13}C$ determinations and they showed unequivalent results. In fact, these two methods are concerned only indirectly with the complete metabolism of exogenous glucose; other phenomena, mainly involving storage processes, are implicated and can interfere to falsify conclusions about glucose transformation into CO_2.

It is evident, then, that this new method of investigation is very useful in the so-important field of glucose metabolism. Experiments have already been carried out or are in progress to compare glucose oxidation in normal, obese or diabetic subjects, to study the effect of drugs (*e.g.* oral antidiabetics) on this process, or to measure the metabolism of various substrates (sugars, oils, alcohols, etc.) at rest or in physical exercise[10-12].

REFERENCES

1 B. Smith and S. Epstein, *Plant Physiol.*, 47 (1971) 380.
2 F. Mosora, M. Lacroix and J. Duchesne, *C. R. Acad. Sci. Ser. D*, 273 (1971) 1423.
3 M. Lacroix and F. Mosora, in *Isotope Ratios as Pollutant Source and Behaviour Indicators*, Int. At. En. Ag. Vienna, 1975, p. 343.
4 M. Lacroix, F. Mosora, M. Pontus, P. Lefebvre, A. Luyckx and G. Lopez Habib, *Science*, 181 (1973) 445.

5 F. Mosora, M. Lacroix, M. Pontus and J. Duchesne, *Bull. Cl. Sci. Acad. Roy. Belg.*, 58 (1972) 565.
6 H. Craig, *Geochim. Cosmochim. Acta*, 12 (1957) 133.
7 D.L. Costill, A. Benett, G. Branam and D. Eddy, *J. Appl. Physiol.*, 34 (1973) 764.
8 W. Shreeve, in *Proceedings 1st International Conference Stable Isotopes*, Argonne, 1973 (USAEC Conf. 730525), p. 390.
9 P. Felig, J. Wahren and R. Hendler, *Diabetes*, 24 (1975) 468.
10 P. Lefebvre, F. Mosora, M. Lacroix, A. Luyckx, G. Lopez Habib and J. Duchesne, *Diabetes*, 24 (1975) 185.
11 F. Mosora, P. Lefebvre, F. Pirnay, M. Lacroix, A. Luyckx and J. Duchesne, submitted for publication.
12 F. Pirnay, P. Lefebvre, M. Lacroix, A. Luyckx and F. Mosora, in preparation.

MASS FRAGMENTOGRAPHY AS A TOOL FOR ENZYME ACTIVITY ASSAY: MEASUREMENT OF GLUTAMATE DECARBOXYLASE

F. CATTABENI, C.L. GALLI, L. DE ANGELIS and A. MAGGI

Institute of Pharmacology and Pharmacognosy, University of Milan, 20129 Milan (Italy)

SUMMARY

The possibility of using mass fragmentography for the direct measurement of enzymic activities in small brain areas was investigated. In principle, tissue homogenates are incubated with a deuterated precursor and the deuterated product formed is measured selectively, even in presence of high amounts of its endogenous protium form.

We have applied this approach to the measurement of the activity of glutamate decarboxylase (GAD) (E.C. 4.1.1.15). This important enzyme catalyzes the formation of γ-aminobutyric acid (GABA) from glutamic acid in neuronal tissues. GABA and GAD have been implicated as biochemical factors in the pathogenesis of various neurological disorders (epilepsy, Huntington's chorea, Parkinson's disease).

GAD activity is currently measured, by an indirect method, based on the measurement of $^{14}CO_2$ developed by decarboxylation of glutamic acid-1-^{14}C, described by Roberts and Simonsen.

The incubation conditions we use are essentially those described by Roberts and Simonsen, except that glutamic acid-D_5 is used as substrate and a selective inhibitor of GABA transaminase (ethanolamine-O-sulfate; Fowler and John, *Biochem. J.*, 130 (1972) 569) is added.

GABA-D_5 so formed is measured with a mass fragmentographic procedure recently developed by Cattabeni *et al.*, with 5-aminovaleric acid as internal standard. The advantages of this direct method for the measurement of GAD activity over the radioisotopic method will be discussed.

INTRODUCTION

The availability of sensitive and specific assays for the neurotransmitters and their synthesizing enzymes is essential if more is to be learned about the organization of the central nervous system at the synaptic level.

Mass fragmentography is certainly one of the most sensitive and specific techniques successfully applied to the quantitation of several neurotransmitters such as acetylcholine, noradrenaline, dopamine, 5-hydroxytryptamine and γ-aminobutyric acid (GABA)[1-4]. These methods have allowed us to study the functional interplay of the different neurotransmitters in discrete brain areas weighing only few milligrams[5]. Mass fragmentography is also currently applied to the measurement of the turnover rate of neurotransmitters by measuring their specific activity at different times after infusion of a precursor labeled with deuterium[6].

The measurement of the activity of the enzymes responsible for the formation of the neurotransmitters in the nerve endings is achieved mainly through radioisotopic methods[7-9]. For example, glutamate decarboxylase (GAD), the enzyme that catalyzes the formation of GABA from glutamic acid, is assayed by measuring the $^{14}CO_2$ formed from L-[1-^{14}C]-glutamate[8,10]. However, this assay is not specific for this enzyme, since there are several pathways by which tissue homogenates catabolize glutamate. In fact it has been concluded that the measurement of $^{14}CO_2$ evolution is not a valid estimate of true GAD activity[11,12]. Therefore, a non-isotopic method has been recently described, which is based on the spectrophotofluorimetric measurement of GABA formed from glutamate[13].

This approach, although more specific, has the disadvantage that it is less sensitive than the isotopic one and is rather laborious, since it is always necessary to run parallel samples in order to subtract the amount of endogenous GABA from the total GABA measured after the incubation.

Mass fragmentography can easily solve both problems, *i.e.* sensitivity and specificity of the assay. In fact it is possible to incubate tissue homogenates with a deuterated precursor and measure, with the high sensitivity intrinsic to mass fragmentography, the deuterated product formed, even in presence of high amounts of its endogenous protium form.

This new approach for enzymic activity measurement has been applied to the assay of GAD, by using glutamate-D_5 as substrate and measuring, with a recently developed mass fragmentographic method[4], the deuterated GABA formed. The first results obtained are reported here.

MATERIALS AND METHODS

The cerebellum of male Sprague-Dawley rats (125-150 g, Charles River), killed by guillotine, is excised and homogenized in 4 volumes of water. 50 μl of the homogenate are added to 50 μl of 80 m*M* phosphate buffer (pH 6.4), containing 2 m*M* glutamate-D_5 (Merck, Sharp and Dohme, Canada) and 1 m*M* pyridoxal phosphate and incubated at 37° for the appropriate time. The enzymic reaction is stopped by boiling for 2.5 min. Next, 100 μl of 1*M* formic acid are added and the mixture is centrifuged at 20,000 *g* for 15 min. An aliquot of the supernatant (20 μl) is transferred into a screw-cap vial and the internal standard 5-amino-valeric acid (5-AVA) is added (about 1 nmole). The solution is then dried under reduced pressure, and 30 μl of a 3:1 mixture of Sil-Prep (Applied Science, State College, Pa., U.S.A.) and BSTFA (Pierce, Rockford, Ill., U.S.A.) is added[4]. After at least 2 h, an aliquot of the reaction mixture is injected into the gas chromatograph-mass spectrometer (Finnigan 3000 F, equipped with a six--channel PROMIM) focused on *m/e* 174.304 for GABA, 175.309 for GABA-D_5 and 174.318 for 5-AVA.

Gas chromatography-mass spectrometry conditions: column 3 m, silanized glass, packed with 3% OV-17. Temperatures: oven 190°, flash heater 220°, separator 250°. Helium flow 30 ml/min. Electron energy 70 eV. Beam current 0.21 mA. Multiplier voltage 2.00 kV.

RESULTS

Fig. 1 shows the mass spectra of the trimethylsilyl derivative for glutamic acid protium form and D_5. The most prominent peaks are at *m/e* 246 and 251 respectively, corresponding to the loss of the -COOTMS group from the molecular ion. As can be seen, the glutamate-D_5 did not contain any glutamate protium form, and only a small percentage of D_4 and D_3 were present.

After incubation of glutamate-D_5 in homogenates of rat cerebellum, deuterated GABA was formed by decarboxylation of the carboxyl in α position with respect to the amino group. Therefore, the deuterated GABA formed contained 5 deuterium atoms: two on each of the two carbon atoms in α and β positions with respect to the carboxyl group and one on the γ-carbon atom. Since GABA-D_5 is not commercially available, its mass spectrum was extrapolated from that of the trimethylsilyl derivative of GABA protium form (Fig. 2).

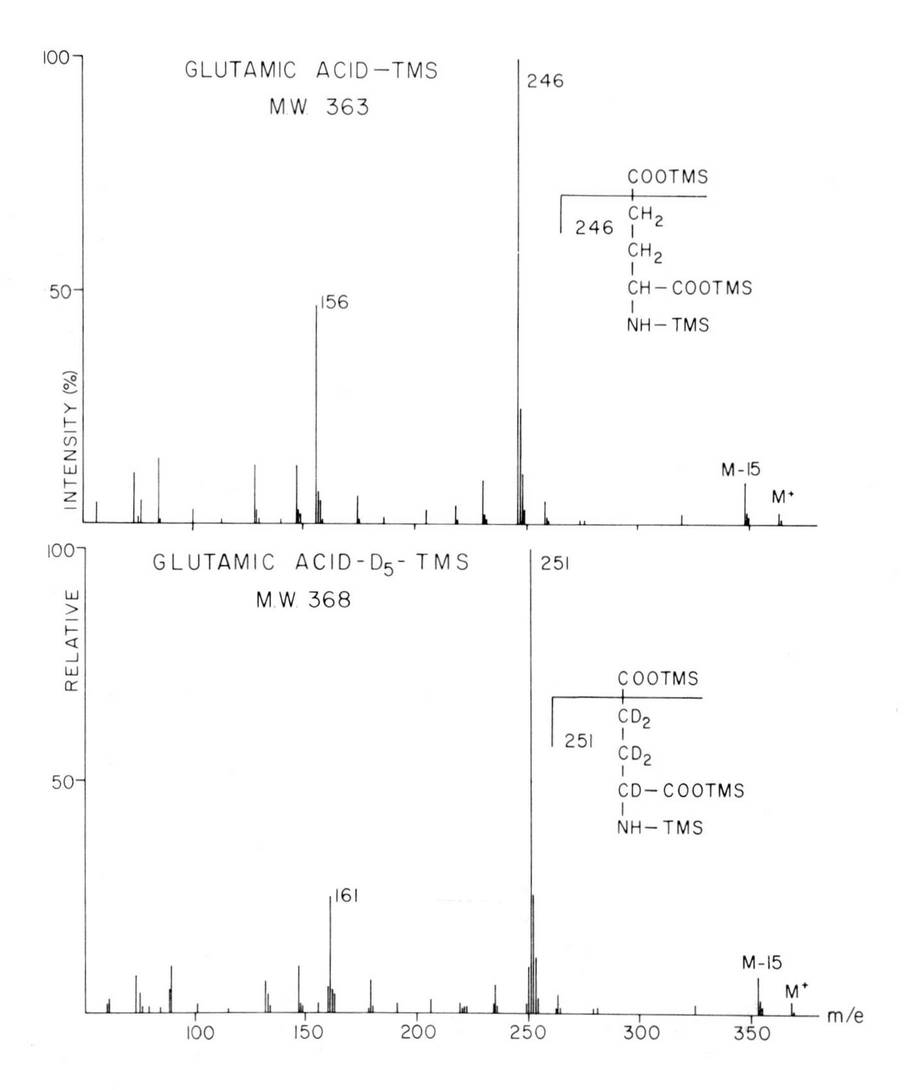

Fig. 1. Mass spectra of the trimethyl-silyl derivatives of glutamic acid and glutamic acid-D_5.

For GABA-D_5, the M-15 ion was expected to have *m/e* 309, and the ion corresponding to that with *m/e* 174 for GABA protium form should be at *m/e* 175. Moreover, the ion density ratio for the fragments with *m/e* 174 and 304, and 175 and 309, for GABA and GABA-D_5 respectively, should have the same value. Since the values found are indeed coincident (4.60 $\pm$ 0.05 for GABA and 4.65 $\pm$ 0.07 for GABA-D_5), it is possible to conclude that the

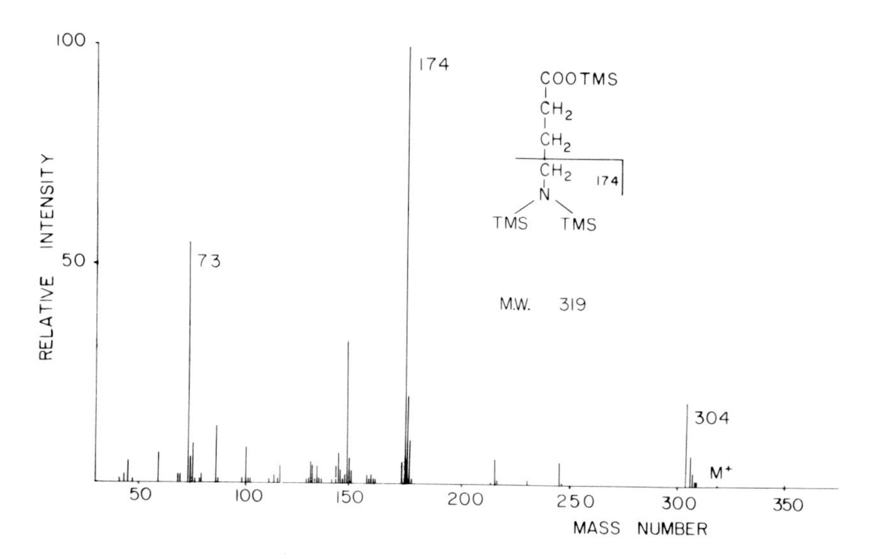

Fig. 2. Mass spectrum of the trimethyl-silyl derivative of GABA.

compound, eluting at the same retention time as that of GABA in supernatants of rat cerebellar homogenates incubated with glutamate-D_5, is GABA-D_5. Moreover, the decarboxylation reaction is presumed to be enzymic, since the same incubation mixture, boiled for 2.5 min before the addition of the substrate, completely prevented the formation of deuterated GABA (Fig. 3).

Conversely, the addition of pyridoxal phosphate to the incubation mixture, increased the formation of deuterated GABA. This is in accord with the known characteristics of GAD present in neuronal tissue, which requires pyridoxal phosphate as cofactor[14].

Fig. 4 shows that production of GABA-D_5 from glutamate-D_5 was linear between 0 and 30 min. A plateau was reached, in our conditions, after about 45 min.

DISCUSSION

We have investigated the possibility of using mass fragmentography for enzymic activity measurements. In principle, tissue homogenates are incubated with a deuterated precursor and the deuterated product formed is measured selectively, even in presence of high amounts of its endogenous protium form.

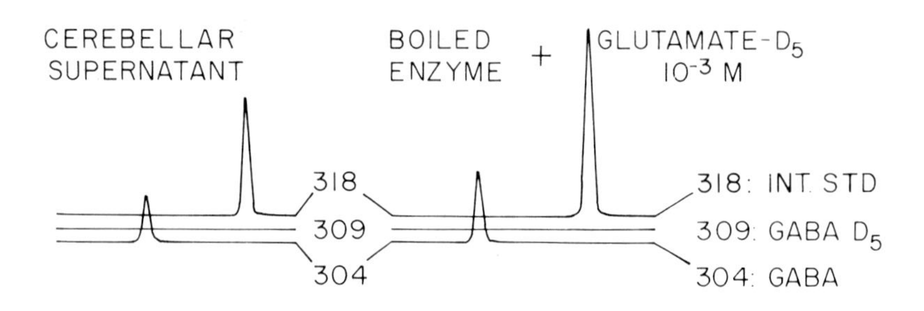

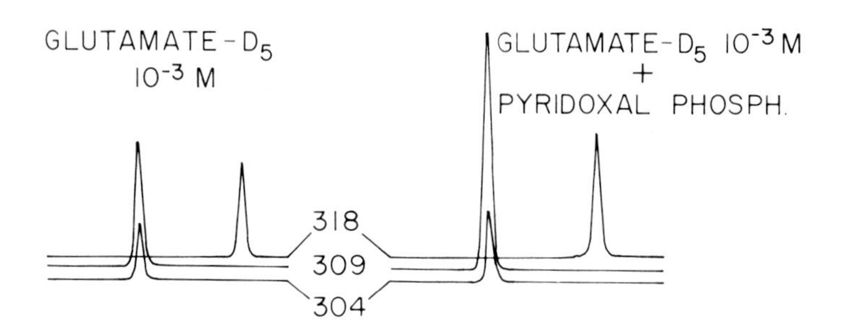

Fig. 3. Mass fragmentograms obtained by analyzing GABA and GABA-D_5 formed after incubating cerebellar homogenates with glutamate-D_5, 10^{-3} *M*.

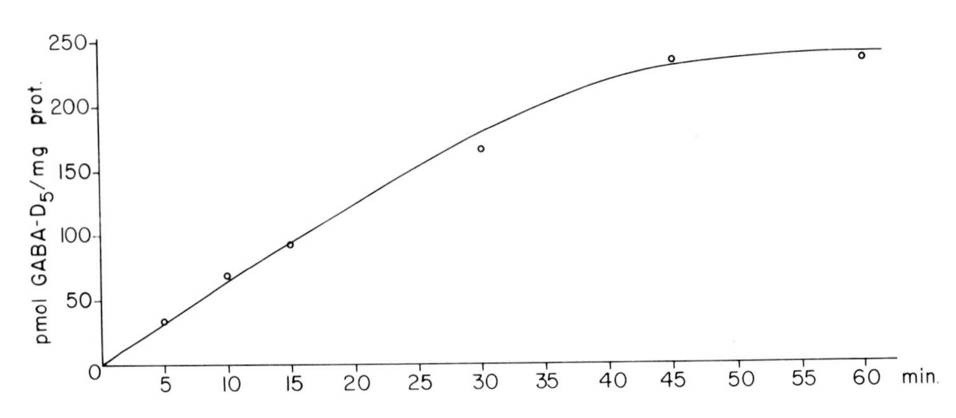

Fig. 4. GABA-D_5 formed from glutamate-D_5 as a function of time of incubation (for conditions, see Materials and methods).

We have applied this approach to the measurement of GAD, the enzyme synthesizing GABA, a major inhibitory neurotransmitter in mammalian brain. The substrate used is glutamate-D_5 and the product formed is GABA-D_5, as can be deduced from the retention time, coincident with that of GABA (Fig. 3), the presence of the fragments at *m/e* 309 (Fig. 3) and 175, and the ratio of the ion densities of these fragments, which is identical with that obtained from the corresponding fragments of GABA protium form. Since the formation of GABA-D_5 is enzyme dependent (Fig. 3) and linear with time (Fig. 4), we can conclude that GAD activity can be effectively measured by this mass fragmentographic method.

This approach not only has a sensitivity comparable to that obtained with the radioisotopic method - GAD can be easily measured in one milligram of brain tissue - but is much more specific, since it is based on the measurement of GABA formed rather than on the measurement of CO_2 evolved from glutamate. In fact it has been reported that CO_2 can derive from glutamate metabolized by alternative pathways[11,15]. However, before this method is applied for routine measurements of GAD, it has to be investigated whether the enzyme discriminates between glutamate-D_5 and the protium form. Work is now in progress in our laboratory to clarify whether this isotopic effect is detectable in our experimental conditions.

ACKNOWLEDGEMENTS

C.L. Galli is the recipient of a fellowship from the Swiss National Research Fund, Commission of Italian Switzerland.

REFERENCES

1 I. Hanin and J. Schubert, *J. Neurochem.*, 23 (1974) 819.
2 S.H. Koslow, F. Cattabeni and E. Costa, *Science*, 176 (1972) 177.
3 F. Cattabeni, S.H. Koslow and E. Costa, *Science*, 178 (1972) 166.
4 F. Cattabeni, C.L. Galli and T. Eros, *Anal. Biochem.*, (1976) in press.
5 S.H. Koslow, G. Racagni and E. Costa, *Neuropharmacol.*, *13 (1974) 1123.*
6 E. Costa, D.L. Cheney, G. Racagni and G. Zsilla, *Life Sci.*, 17 (1975) 1.
7 T. Nagatsu, M. Levitt and S. Udenfriend, *Anal. Biochem.*, 9 (1964) 122.
8 R.W. Albers and R.O. Brady, *J. Biol. Chem.*, 234 (1959) 926.
9 B.K. Schrier and L. Shuster, *J. Neurochem.*, 14 (1967) 977.
10 E. Roberts and D.G. Simonsen, *Biochem. Pharmacol.*, 12 (1963) 113.
11 R.J. Drummond and A.T. Phillips, *J. Neurochem.*, 23 (1974) 1207.
12 D.L. Martin and L.P. Miller, in E. Roberts, T.N. Chase and D.B. Tower (Editors), *GABA in Nervous System Function*, Raven Press, New York, 1976, pp. 57-58.
13 P. MacDonnell and O. Greengard, *J. Neurochem.*, 24 (1975) 615.

14 J.Y. Wu, in E. Roberts, T.N. Chase and D.B. Tower (Editors), *GABA in Nervous System Function*, Raven Press, New York, 1976, pp. 7-55.
15 S.H. Wilson, B.K. Schrier, J.L. Farber, E.J. Thompson, R.N. Rosenberg, A.J. Blume and M.W. Nirenberg, *J. Biol. Chem.*, 247 (1973) 3159.

QUANTITATIVE DETERMINATION OF PIPERIDINE IN THE RAT BRAIN BY SELECTED ION MONITORING FOLLOWING DINITROPHENYL DERIVATIZATION

I. MATSUMOTO and T. SHINKA

Department of Biochemistry, Kurume University School of Medicine, Kurume (Japan)

and

Y. KASE and Y. OKANO

Department of Chemical Pharmacology, Faculty of Pharmaceutical Sciences, Kumamoto University, Kumamoto (Japan)

SUMMARY

In the studies dealing with biogenic amines and related compounds, analytical methods were required to be strictly sensitive and specific. Like biogenic amines, piperidine has been known through producing potent nicotine-like actions on the peripheral and central nervous systems, and it is present in trace amount in the brains of many species including human.

We have studied the quantitative analysis of piperidine in biological materials and developed an analytical method for trace amounts of piperidine. Seiler and Schneider recently reported a method for the chromatographic separation and mass spectrometric determination of piperidine from biological samples. We have independently developed a new method of mass fragmentography of dinitrophenyl derivatives using the deuterated analogues as internal standard in absolute amounts as low as 10^{-11} moles. Experimental conditions, precision and sensitivity are discussed.

INTRODUCTION

Piperidine has potent nicotine-like actions on the peripheral[1-3] and central nervous system (CNS)[4,5]. Like nicotine, it acts on the autonomic ganglia, neuromuscular junctions and chemoreceptors, *i.e.* it elicits stimulation in small doses and depression in large doses. In the CNS, it also produces synaptic stimulation followed by depression[5]. In addition, it exerts tranquilizing effects on schizophrenic patients[6] and counteracts experimentally induced aggressiveness in mice and rats[5]. The findings obtained from chemical stimulation of the brain by piperidine suggest that the amine itself affects neural mechanisms serving regulation of emotional behaviour, sleep and extrapyramidal functions[7]. Furthermore, piperidine can mimic the action of acetylcholine on synaptic sites in the brain of mammals [5] and in the cerebral ganglia of molluscs[8].

Piperidine is a normal constituent of mammalian brain[9,10] and urine [10,11], of human cerebrospinal fluid[12] and of cerebral ganglia of molluscs [13]. Piperidine can be produced by the action of a decarboxylase in brain tissue *in vivo*[14] as well as *in vitro*[15] from pipecolic acid, which is an intermediate of lysine metabolism and occurs in serum[16], urine[17] and the brain[18] of mammals. The amine and its precursor are not ubiquitously distributed in the brain but are concentrated in some regions[19,20]. This suggests that piperidine may play an important functional role in the brain. It is, therefore, of interest to determine the concentration of piperidine in various regions of the mammalian brain and its correlation with physiological functions.

Several methods for quantitative determination of piperidine have been reported[10,19,21], but they seem inadequate in sensitivity and specificity. Recently, Seiler and Schneider[22] have introduced a method for quantitation of picomoles of piperidine by the use of a combination of thin-layer chromatography (TLC) and mass spectrometry following 5-di-*n*-butylamino-naphthalene-1-sulphonyl (BANS) derivatization. Barsuhn[23] has also succeeded in quantitation of piperidine of the same order as shown by Seiler and Schneider by using 3,5-dinitrobenzotrifluoride derivatization followed by gas chromatography with electron-capture detection.

The present paper deals with a new highly specific and sensitive method for quantitative determination of piperidine in the rat brain by the use of dinitrophenyl derivatization followed by a selected ion-monitoring technique (SIM), with the aid of deuterium-labelled piperidine as an internal standard.

EXPERIMENTAL

Materials

2,4-Dinitrofluorobenzene (DNFB) was purchased from E. Merck and piperidine-d_{11} (base) was obtained from Merck, Sharp and Dohme, Canada. Other reagents used were of analytical grade and obtained from commercial sources.

Fortly-five male Wister albino rats, weighing about 180 g and 7 weeks old, were used throughout this study.

Extraction and dinitrophenyl derivatization of piperidine

After decapitation of the rats, the brains were removed within 3 min in a cold chamber and placed in ice. About 5 g of brain tissue (correspon-ding to three brains) were homogenized in 20 ml of 5% trichloroacetic acid solution (TCA) with a glass homogenizer. The homogenate was centrifuged at 17,000 *g* for 20 min at 0°. The supernatant was decanted into a glass tube. The precipitate was dissolved in 12 ml of 5% TCA and centrifiged at 17,000 *g* for 20 min. The total supernatants were adjusted to pH 10 with 1 *M* bi-carbonate buffer and carefully steam distilled to separate volatile amines. The distillate (200 ml) was trapped in 1 *M* HCl, concentrated under reduced pressure, followed by neutralization with 0.2 *M* NaOH. Piperidine and other volatile amines in the solution (20 ml) were converted to the corresponding dinitrophenyl (DNP) derivatives as follows.

DNFB 1.3 ml of ethanol solution (500 µmoles in 5.0 ml), was added to the solution. The mixture was heated at 60° for 20 min under continuous shaking. To decompose the excess of DNFB, 3 ml of 0.2 *M* NaOH were added to the mixture, which was again heated at 60° for 60 min. After the mixture had been cooled to room temperature, 15 ml of cyclohexane were added and the mixture was shaken for 15 min. The cyclohexane layer was evaporated. The residue was dissolved in 100 µl ethanol, and 1.0-3.0 µl of the solution were applied to gas chromatography with mass spectrometry (JEOL GC-MS D-100).

Piperidine-d_{11} was added to each of the brain homogenates (final concentration, 5×10^{-8} moles).

Selected ion monitoring (SIM)

The glass column (1 m x 2 mm), packed with 1.5% OV-17 on Shimalite W (80-100 mesh), was used. Helium was used as a carrier gas at a flow-rate of 30 ml/min. The temperatures of column and injection port were 200 and 250°, respectively, and that of the ion source was about 250°. The electron energy was set at 75 eV and the trap current at 300 μA. The ions used for SIM were *m/e* 234 (M-OH) for DNP-piperidine and *m/e* 243 $(M\text{-}OD)^+$ for DNP-piperidine-d_{10}.

RESULTS

Fig. 1 shows spectra of DNP-piperidine and DNP-piperidine-d_{10}. The base peak of the former was *m/e* 234 $(M\text{-}OH)^+$ and of the latter was *m/e* 243 $(M\text{-}OD)^+$. These fragment ions were smaller than their molecular ions by 17 and 18 mass units and had the characteristics of piperidine moiety. Therefore, these ions were available for SIM.

The specificity of this method was based on the assumption that there were no contaminations in the gas chromatographic peaks corresponding to DNP-piperidine or DNP-piperidine-d_{10}. To exclude such possibility of contaminations, the gas chromatographic peak of the sample corresponding to DNP-piperidine was analyzed by GC-MS. The mass spectrum was exactly the

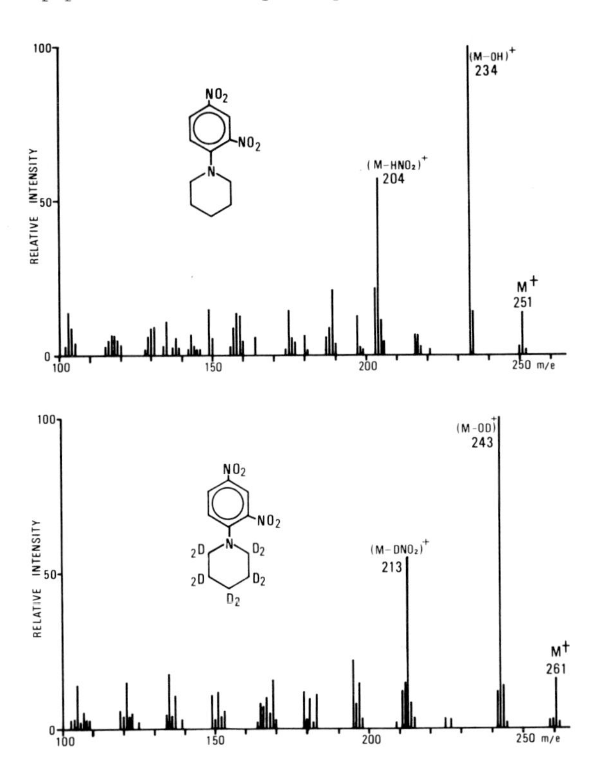

Fig. 1. Mass spectra of unlabelled DNP-piperidine (upper) and DNP-piperidine-d_{10} (lower).

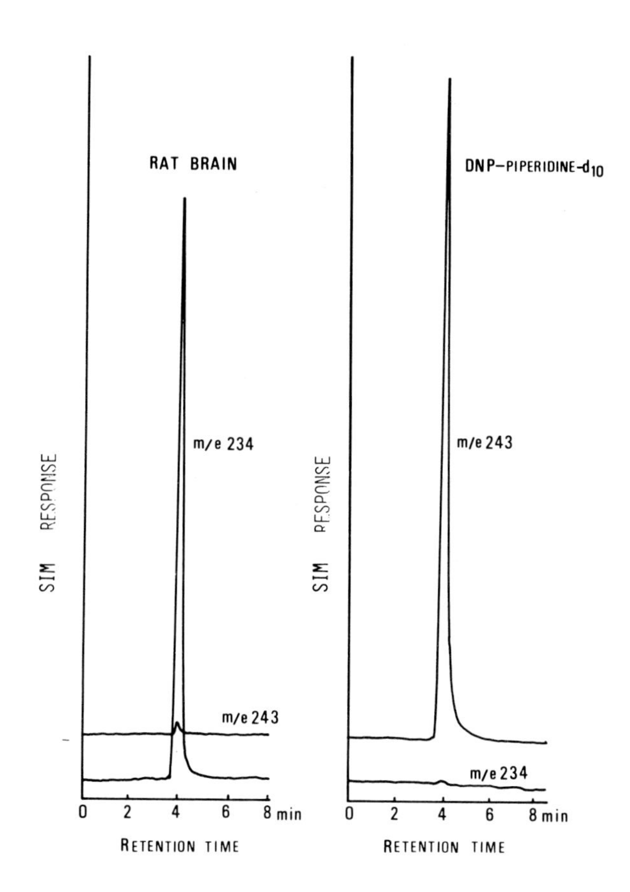

Fig. 2. SIM recordings of the sample (left) and DNP-piperidine-d_{10}. The amplification was 0.5 V for *m/e* 234 and 2 V for *m/e* 243.

same as that of authentic DNP-piperidine. Therefore, no more purification was necessary before gas chromatography.

Fig. 2 shows typical SIM recordings of the sample and of DNP-piperidine-d_{10}. The backgrounds of the sample for selected ion at *m/e* 243 and of DNP-piperidine-d_{10} for selected ion *m/e* 234 were less than 2 and 1.2%, respectively, so they were negligible.

Fig. 3 represents a typical SIM recording and a total ion-monitoring gass chromatogram of the sample mixed with DNP-piperidine-d_{10}.

These data show that this method is apparently quite specific for separation and quantitation of DNP-piperidine from the rat brain.

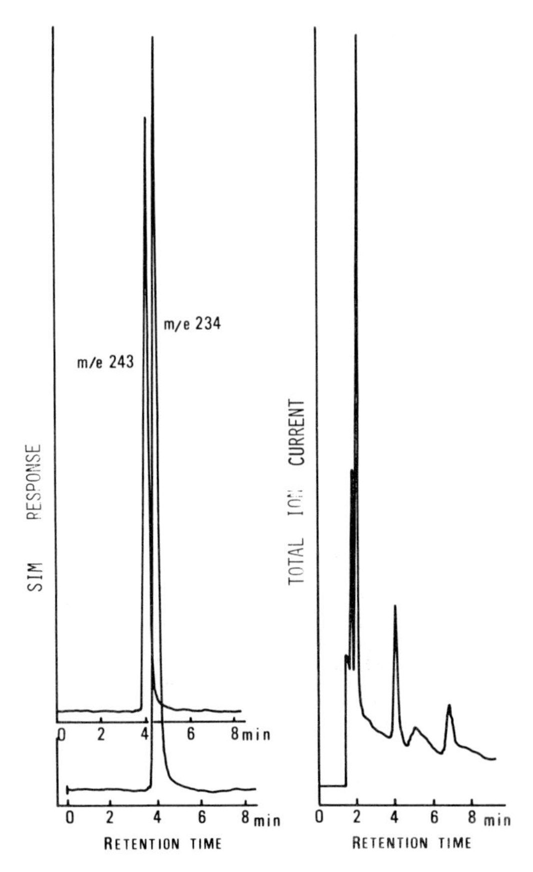

Fig. 3. SIM recording (left) and gas-chromatographic recording (right) of the sample mixed with DNP-piperidine-d_{10}. The amplification was 0.5 V for *m/e* 234 and 2 V for *m/e* 243.

The concentration of DNP-piperidine was calculated from the peak height ratio of *m/e* 234 to *m/e* 243. As shown in Fig. 4, the relative height of the peak at *m/e* 234 against 5×10^{-10} moles of DNP-piperidine-d_{10} was linear in the range of 50 pmoles to 500 pmoles and the lower limit of measurement of DNP-piperidine was as small as 5 pmoles (Fig. 5). Thus, the sensitivity is satisfactory and this method is suitable for determining the nmole quantities of piperidine in the rat brain.

Table I shows the concentration of piperidine in the rat brain determined by this method. The mean value $\pm$ S. D. was 1.81 ± 0.09 nmoles/g wet brain tissue.

Table I

Piperidine concentration in the whole brain of rats

Number of groups	N[+]	Piperidine nmoles/g brain tissue[++]
1	3	1.71
2	3	1.76
3	3	1.91
4	3	1.89
5	3	1.79
Mean ± S.D.		1.81 ± 0.09

[+] N, The number of rat brains.

[++] Mean value for three rat brains.

DISCUSSION

In the method of Seiler and Schneider, BANS-piperidine was isolated by TLC. In general, separation by TLC is time consuming and, furthermore, contamination can hardly be avoided, although Seiler and Schneider expertly handled the problem by converting piperidine into high molecular BANS derivatives, to avoid the noise derived from contaminants. This problem, however, can be settled without difficulty if gas chromatography instead of TLC, is used for the separation of piperidine. In fact, gas chromatographic separation followed by quantitation was successfully applied to 3,5-dinitro--benzotrifluoride derivatives of piperidine by Barsuhn[23].

Gas chromatography is most suitable for the isolation of a compound from contaminants. However, it cannot be used for the identification of the isolated compound. It also takes much time to prepare the sample for TLC. Compared with the above two methods, the SIM technique described here is simpler in sample preparation, more highly specific and as sensitive. By this method, the concentrations of piperidine in the rat brains were found to be 1.71-1.91 nmoles/g tissue, which were about 10-40 times larger than those of Seiler and Schneider[22] and Barsuhn[23].

Such differences may have been due to differences in the internal standard used, that is, Barsuhn used hexamethyleneimine and Seiler and Schneider used pyrrolidine. These two compounds are different from piperidine and, in particular, pyrrolidine also exists in the brain[9]. So, if

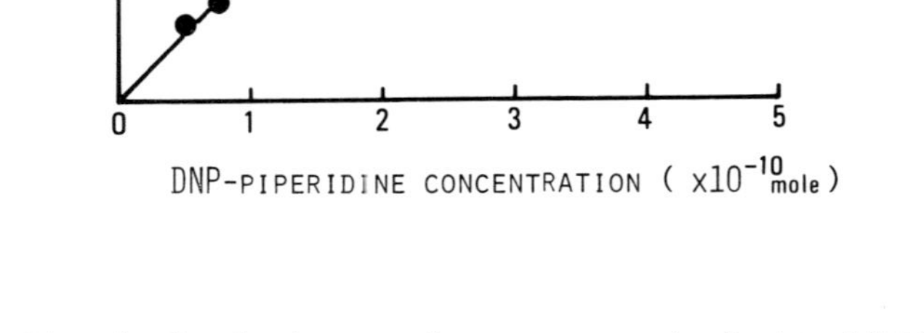

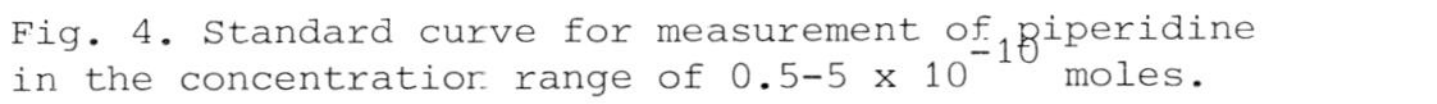
Fig. 4. Standard curve for measurement of piperidine in the concentration range of 0.5-5 x 10^{-10} moles.

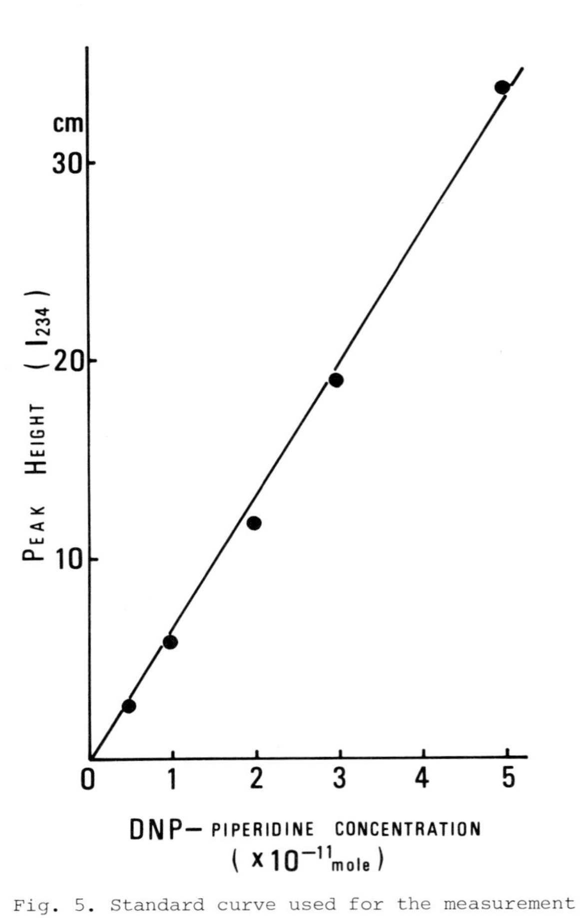

Fig. 5. Standard curve used for the measurement of DNP-piperidine.

pyrrolidine is used as an internal standard, lower values of piperidine will be given. It has now been demonstrated that piperidine can be quantitatively determined with high specificity and sensitivity by using deuterium-labelled piperidine as an internal standard.

REFERENCES

1 U.S. von Euler, *Acta Pharmacol. Toxicol.*, 1 (1945) 29.
2 M.F. Lockett, *Brit. J. Pharmacol.*, 4 (1949) 111.
3 Y. Kasé, T. Miyata and T. Yuizono, *Jap. J. Pharmacol.*, 17 (1967) 475.
4 L.G. Abood, F. Rinaldi and U. Eagleton, *Nature (London)*, 191 (1961) 201.
5 Y. Kasé, T. Miyata, Y. Kamikawa and M. Kataoka, *Jap. J. Pharmacol.*, 19 (1969) 300.
6 D.C. Tasher, L.G. Abood, F.A. Gibbs and E. Gibbs, *J. Neuropsychiat.*, 1 (1960) 266.
7 T. Miyata, K. Kamata, M. Nishikibe, Y. Kasé, K. Takahama and Y. Okano, *Life Sci.*, 15 (1974) 1135.
8 M. Stepita-Klauco, H. Dolezalova and E. Giacobini, *Brain Res.*, 63 (1973) 141.
9 C.G. Honegger and F. Honegger, *Nature (London)*, 185 (1960) 530.
10 Y. Kasé, M. Kataoka and T. Miyata, *Jap. J. Pharmacol.*, 19 (1969) 354.
11 U.S. von Euler, *Nature (London)*, 154 (1944) 17.
12 T.L. Perry, S. Hansen and L.C. Jenkins, *J. Neurochem.*, 11 (1964) 49.
13 H. Dolezalova, E. Giacobini, N. Seiler and H.H. Schneider, *Brain Res.*, 55 (1973) 242.
14 Y. Kasé, Y. Okano, Y. Yamanishi, M. Kataoka, K. Kitahara and T. Miyata, *Life Sci.*, 9 (1970) 1381.
15 Y. Kasé, M. Kataoka and Miyata, *Life Sci.*, 6 (1967) 2427.
16 T. Yoritaka and C. Hiwaki, *Nagasaki Igakkai Zasshi*, 34 (1959) 277.
17 F. Tominaga and C. Hiwaki, *Nagasaki Igakkai Zasshi*, 33 (1957) 1435.
18 Y. Kasé, M. Kataoka, T. Miyata and Y. Okano, *Life Sci.*, 13 (1973) 867.
19 M. Kataoka, Y. Kasé, T. Miyata and E. Kawahito, *J. Neurochem.*, 17 (1969) 291.
20 T.L. Perry, S.L. Hansen and L. Macdougall, *J. Neurochem.*, 14 (1967) 775.
21 R. Nixon, *Anal. Biochem.*, 48 (1972) 460.
22 N. Seiler and H.H. Schneider, *Biomed. Mass Spectrom.*, 1 (1974) 381.
23 C. Barsuhn, *Life Sci.*, 18 (1976) 419.